Chemical Admixtures for Concrete

Chemical Admixtures for Concrete

Second Edition

M.R. Rixom and N.P. Mailvaganam

London New York
E. & F.N. SPON

First published in 1978 by
E. & F.N. Spon Ltd
11 New Fetter Lane, London EC4P 4EE

Second Edition published in 1986 by
E. & F.N. Spon Ltd
11 New Fetter Lane, London EC4P 4EE
Published in the USA by
E. & F.N. Spon
29 West 35th Street, New York NY 10001

Printed in Great Britain at
the University Press, Cambridge

ISBN 0 419 12630 9

British Library Cataloguing in Publication Data
Rixom, M.R.
Chemical admixtures for concrete.——2nd ed.
1. Concrete——Additives
1. Title II. Mailvaganam, N.P.
620.1'36 TP884.A3
ISBN 0-419-12630-9

Library of Congress Cataloging-in-Publication Data
Rixom, M. R.
Chemical admixtures for concrete.
Includes bibliographies and index.
1. Concrete——Additives. I. Mailvaganam, N. P.
(Noel, P.), 1938- . II. Title.
TP881.R56 1986 666'.893 85-14474
ISBN 0-419-12630-9

Contents

Acknowledgements

We would like to acknowledge the contributions made during the preparation of this second edition and in particular the following people and organizations which gave permission for material to be utilized:

John Mowlem Ltd . Patterson-Candy International Ltd . Sunderland and South Shields Area Water Authority . John Laing Construction Ltd . Sir Alfred McAlpine Ltd . Howard-Doris Ltd . Taylor Woodrow International Ltd . Barber Greene. . R.J. Shutz . W.A. LaFrangh . British Lift Slab Ltd . RMC Transite Ltd . RMC (Thames Valley) Ltd . Anglian Building Products Ltd . Tallaght Block Ltd . Reema Concrete Ltd . Protex Industries . Master Builders Co . SKW (Trostberg) GmbH . MAC S.p.a. . Flowmix Canada Cement Lafarge . W.R. Grace . C. Maurice Co (Contractors) Ltd . Clondalkin Concrete Ltd . New Brunswick Power Commission . Pergamon Press . The Concrete Society . The Portland Cement Association . American Society for Testing and Materials . Matériaux et Constructions . RILEM . The American Ceramic Society . The American Concrete Institute . The Institute of Construction and Architecture of the Slovak Academy of Sciences.

In addition, those readers who attended the Concrete International meeting in 1980 will recognize Chapter Six as an updated version of the excellent paper presented by Mr T. Tipler of the Cement and Concrete Association. We are grateful to him and to the Concrete Society for permission to use his paper as the basis of this chapter.

Finally we would like to thank our companies Cormix Ltd and Sternson Ltd for allowing us to publish the book and acknowledge the contribution made by friends and colleagues within our companies.

Foreword to second edition

Since the first edition was published in 1978, several changes have occurred which the original author considered necessitated a reappraisal of the book's contents leading to an enlarged and revised edition. Apart from a general updating of the literature, and rectifying minor errors and deficiencies, the following observations influenced the decision to produce a further edition:

(a) The use of admixtures, as expected, has gained greater acceptance so that the amount of concrete which contains admixtures has increased. This has led to a better understanding of the properties of admixtures by the concrete engineer and technologist, so that he is less dependent on the reliability of the advice of the admixture supplier and has become a more discerning specifier and user. On the other hand, the widespread problems of concrete damage caused by earlier indiscriminate use of calcium chloride, sometimes in flake form at high concentrations, has not lessened the awareness of potential dangers that could exist.
Co-operation between industry and governmental agencies has produced new or modified standard testing methods allowing limits to be set for minimum levels of acceptability, and providing a means by which products can be fairly evaluated on a comparative basis. In certain countries this has been further developed to certification of a product's fitness for use, without which sale of such materials is not allowed.
In view of this it has been decided to include a chapter on the standards and controls that exist in various countries. This has replaced the original Chapter 6 which has been omitted from this edition (see [d] below).

(b) The construction industry has continued to become more international; as the economic climate of the construction industry has declined in most western countries, so has the need to be involved in overseas contracts become more important. Many more engineers are finding themselves grappling with products, standards, and specifications which emanate from other countries, and working in climatic conditions which make admixtures not only desirable, but necessary. In view of this, apart from the chapter on standards etc., more information has been given on the influence of high temperatures on concrete which contains admixtures.
(c) At the time of writing the first edition, the majority of the applicational information was based on UK practice and experience. The international response to the book has been such that it was felt desirable to introduce a more international flavour to the applications chapter. In addition, it could not be forgotten that the USA represents the largest single market for chemical admixtures and the efforts of the American admixture industry, the construction industry, and the ASTM have contributed more to the present state of the art than any other single nation. It was considered timely, therefore, to introduce a co-author, Noel Mailvaganam, who has experience of the situation in America. This was particularly welcomed by the original author, Roger Rixom, who has known Noel Mailvaganam over many years, and has resulted in an expansion of Chapter 5 leading, it is believed, to a much more authoritative and useful section.
(d) It was felt that neither of the authors of the revised edition could contribute further to the chapter on the analysis of concrete for admixture content as this lay outside the field of their experiences. On the other hand, the information presented in the original edition is now largely out of date; for this reason this chapter has been omitted from the new edition. It is to be hoped that those qualified in this field could be persuaded to produce a book on this important topic, or perhaps to contribute a chapter in a subsequent edition of this book.

The authors hope that the additions and changes described above will result in a book both gaining a wider readership and finding a greater usefulness to the individual reader. Finally, the authors would like to thank all those who have constructively criticised the original edition.

Introduction

Since the publication of the first edition of this book, most major industrial countries have seen recessional trends in their construction industries, the hardest hit being the large civil engineering contracts. Against this background, the chemical admixture business has shown remarkable growth worldwide and has become a major chemical business with an annual turnover in the region of US $600 million.

A typical example of the resilience of the industry to withstand a declining or stagnating construction industry is shown in Fig. 0.1 where the cement use and admixture production in Western Germany over the period 1978 - 1981 are compared.

It is appreciated that the admixture production figures contain an export element but, nevertheless, the health of the admixture business in comparison to the industrial climate within which it works is apparent. The reasons for this apparent depression-proof nature of the admixture business are believed to be:

(a) In those countries where the use of admixtures falls well below the potential use, there is considerable opportunity for the degree of penetration to be increased, resulting in a net growth. Indeed, in recessionary times with reduced concrete volumes, there is commercial pressure to improve margins and the use of admixtures is often part of a cost-saving strategy.
(b) In countries where penetration is already high, growth has come largely from the use of superplasticizers which are added at much higher dosage levels than traditional water-reducing agents. In this

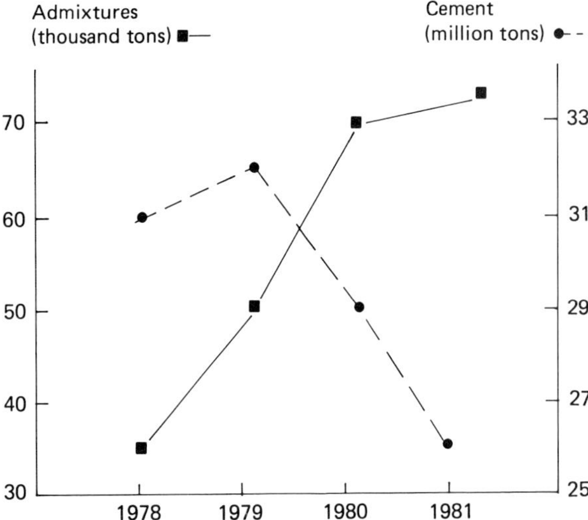

Fig. 0.1

area there has also been much technical development, particularly with regard to products based on polymerized naphthalene sulphonates. It is the view of the authors that future growth over the next five years will come in the main from the area of superplasticizers.

(c) The continued movement of the construction industry to involvement in overseas projects has called for admixtures to assist concreting in extremes of climatic conditions. Concrete oil platforms and Middle East projects are two such examples.

A picture of the global development of cement and admixture usage over the decade 1971 – 1981 is given in Table 0.1.

The increase in the use of admixtures has been reflected in a greater interest in admixture materials and technology from all sectors of the industry. Thus all types and levels of information are demanded, including the simple, but very important, procedures for use on site, the more complex requirements of the design engineer regarding structural characteristics of concrete containing admixtures.

In addition, the cement chemist, the specifier of concrete, the student, the admixture supplier, and many others, require detailed or general information on some aspect of admixtures. The purpose of this book, therefore, is to compile into one volume the available data on admixtures in order to satisfy the demands of all interested parties. It is appreciated that this ambitious objective is unlikely to be achieved to every reader's satisfaction and, to keep the project down to a manageable proportion, it

Table 0.1

Geographical area	Estimated cement consumption (million tons)		Estimated* admixture consumption ('000 tons)	
	1971	1981	1971	1981
USA	70	70	200	250
South America	40	60	40	40
Western Europe	155	160	100	220
Africa	25	40	10	15
Middle East	N/A	35	10	50
Far East	105	140	80	165
Australia	5	6	5	15

*excluding calcium chloride accelerators

has been decided to include only those categories of admixtures which are currently used to any significant degree. Also excluded are those materials, such as pozzolans and pigments, which, although classified by other workers as admixtures, are outside both the normal consideration of those products and the authors' own sphere of interest and knowledge.

The types of admixtures considered in this book are: water-reducing agents (including superplasticizers and retarders), air-entraining agent, waterproofers, and accelerators. Within each product category, the following aspects will be covered: background and definition, chemistry, effects on the cement-water system, effects on plastic concrete, effects on hardened concrete (engineering and durability aspects).

The practical use of admixtures in a variety of applications in the site-batched, ready-mixed, and precast industry is covered in a separate chapter where, apart from *actual examples of use, factors such as storage, safety, and dispensing equipment are discussed.* A short chapter is devoted to international standards and specifications.

It is hoped that in some small way this book will add to an already much improved situation by providing a source of information for those engaged in the manufacture, use and specification of concrete.

1 Water-reducing agents

1.1 Background and definitions

The water-reducing admixtures are the group of products which possess as their primary function the ability to produce concrete of a given workability, as measured by slump or compacting factor, at a lower water–cement ratio than that of a control concrete containing no admixture.

The earliest known published reference to the use of small amounts of organic materials to increase the fluidity of cement containing compositions, was made in 1932 [1] where polymerized naphthalene formaldehyde sulphonate salts were claimed as useful in this role. This was followed during the mid 1930s to early 1940s by numerous disclosures regarding the use of lignosulphonates and improved compositions [2-9].

The lignosulphonates formed the basis of almost all the available water-reducing admixtures until the 1950s when the hydroxycarboxylic acid salts were developed which have grown to occupy a significant but, nevertheless, still a minority position in this product group. Materials such as glucose and hydroxylated polymers obtained by the partial hydrolysis of polysaccharides have been widely used in North America. The polymers usually have a low molecular weight and contain glycoside units ranging from 3–25. In addition, other chemical and admixture types have been included into the water-reducing admixtures formulations to produce five types within this category.

The *normal* water-reducing admixtures allow a reduction in the water–cement ratio at a given workability without significantly affecting the setting characteristics of the concrete. In practice, this effect can be utilized in three ways:

(a) By the addition of the admixture with a reduction in the water–cement ratio, a concrete having the same workability as the control concrete can be obtained, with unconfined compressive strengths at all ages which exceed those of the control.
(b) If the admixture is added directly to a concrete as part of the gauging water with no other changes to the mix proportions, a concrete possessing similar strength development characteristics is obtained, yet having a greater workability than the control concrete.
(c) A concrete with similar workability and strength development characteristics can be obtained at lower cement contents than a control concrete without adversely effecting the durability or engineering properties of the concrete.

In all three ways of use, this type of admixture can be regarded as a cement 'saver' as illustrated in Fig. 1.1.

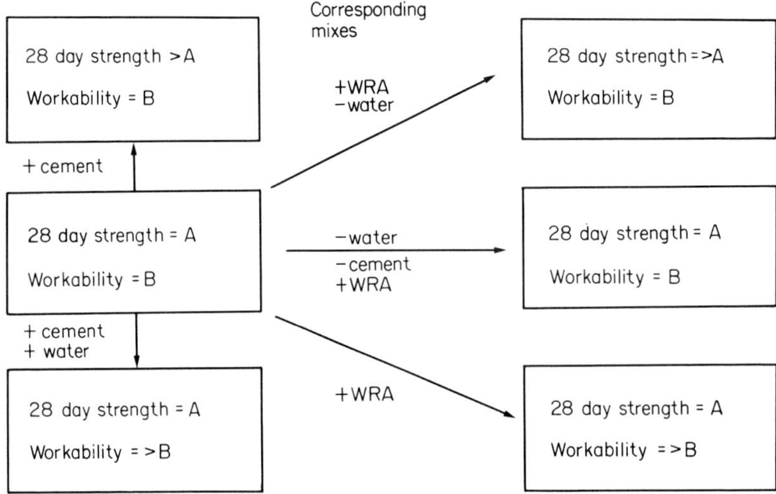

Fig. 1.1 The concept of corresponding mixes.

Corresponding mixes are, therefore, concrete mixes having the same workability and 28-day strength characteristics, but the mix containing the water-reducing admixture will have a lower cement content than the other mix. In practice, of course, the parameters of workability and strength are dictated by the requirements of the particular situation; in areas of high steel content, a high workability will be required, whilst in the production of extruded prestressed lintels, a very low workability is needed. In both cases the strength requirements will be dictated by the load-bearing characteristics of the application. Thus in comparing any properties of admixture-containing concrete, the results of corresponding mixes should be studied, whether the investigation be related to strength, durability factors or statistical considerations, such as standard deviation.

Although the pictorial comparison shown in Fig. 1.1 and discussed above is true at low and average cement contents up to about 350 kg m^{-3}, it is more difficult to obtain higher strengths and workability by further increasing the cement content. It is in this area that the hydroxycarboxylic acid water-reducing admixtures are particularly beneficial, enabling considerable increases in strength to be obtained without the expense and undesirable side effects of large cement increments.

The other members of the water-reducing admixture group possess some other function which could not be obtained by mix design considerations.

The *accelerating* water-reducing admixtures, whilst possessing the water-reducing capability of the 'normal' category, give higher strengths during the earlier hydration period, which is particularly useful at lower temperatures.

The *retarding* water-reducing admixtures again behave in a similar manner to the 'normal' materials and are often of similar chemical composition used at a higher dosage level, but extend the period of time when the concrete is in the workable state. This means that the time available for transport, handling and placing is lengthened. In fact, although a few materials are available which exert only a retarding influence on concrete and have little or no water-reducing capacity, the vast majority (about 95%) of materials called 'retarders' are actually retarding water-reducing admixtures and are considered as such in this book.

The way in which the three types of water-reducing admixtures discussed so far affect the strength gain characteristics of concrete containing them is shown in Fig. 1.2. The four concrete mixes have been designed to have approximately the same 28-day compressive strength, i.e. the admixture-containing mixes would contain approximately 10% less concrete than the control mixes.

The *air-entraining* water-reducing agents possess the ability to entrain microscopic air bubbles into the cement paste whilst allowing a reduction in the water–cement ratio greater than that which would be obtained by the air entrainment itself. They are available in the 'normal' and retarding form and also fall into two types depending on the level of air entrainment; the first type entrains only about 1 to 2% of additional air and is normally used to increase the internal surface of the concrete to redress any deficiencies in fine aggregate gradings. The second type results in concrete containing 3 to 6% of air and is used to enhance the durability of the concrete to freeze–thaw conditions.

The advantages of using this type of material rather than a straight air-entraining agent are based mainly on minimizing the deleterious effect that

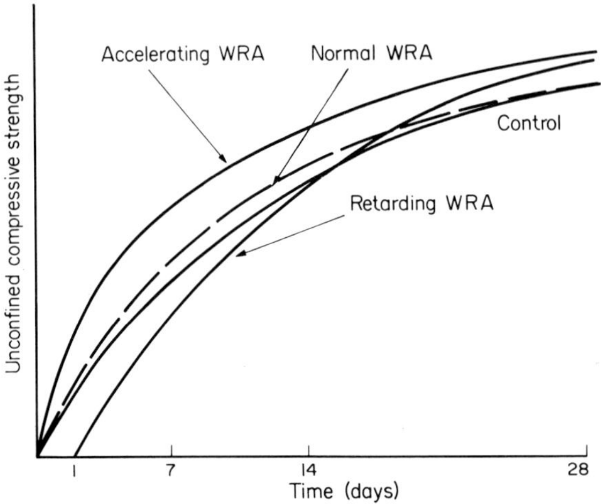

Fig. 1.2 Compressive strength development of concrete containing various types of water-reducing admixture.

air entrainment has on compressive strength, as shown in Fig. 1.3. Thus in a typical concrete mix, up to 3% air can be entrained without any alteration to the mix design or reduction in compressive strength when an air-entraining water-reducing admixture is used.

The *superplasticizers* are an extension of the normal water-reducing admixtures in that they are formulated from materials (see Section 2.1) which allow much greater additions to be made (up to 10 times more than normal water-reducing admixtures) to the concrete mix without producing adverse side effects, such as air entrainment and excessive retardation. In this way concrete can be produced at normal water–cement ratios having such extreme workability that it is almost self-levelling in character.

The extreme workability can also be used to effect considerable reductions in the water–cement ratio, so that very high strengths can be obtained. Fig. 1.4 shows a typical strength development pattern of a superplasticizer-containing mix in comparison to a control concrete of the same workability but containing no admixture [10].

1.2 The chemistry of the water-reducing admixtures

Although the variety of admixtures that are commercially available are marketed under a multitude of benefit orientated classifications, namely waterproofers, workability aids, etc., it is possible to categorize the basic chemicals used in a reasonable way as shown in Table 1.1.

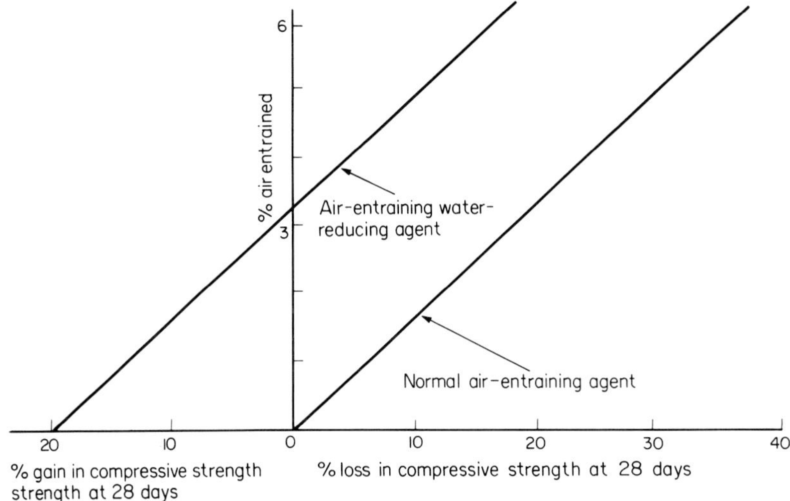

Fig. 1.3 The effect of air entrainment on compressive strength of concrete containing a water-reducing air-entraining agent and a normal air-entraining agent.

It can be seen, therefore, that only five chemical materials form the basis of all the water-reducing admixtures, i.e. lignosulphonate, hydroxy-carboxylic acid, hydroxylated polymers, formaldehyde naphthalene sulphonate, and formaldehyde melamine sulphonate salts.

1.2.1 Lignosulphonates

Lignin is a complex material which forms approximately 20% of the composition of wood. During the process for the production of paper-making pulp from wood, a waste liquor is formed as a by-product containing a complex mixture of substances, including decomposition products of lignin and cellulose, sulphonation products of lignin, various carbohydrates (sugars) and free sulphurous acid or sulphates. Subsequent neutralization, precipitation and fermentation processes [11] produce a range of lignosulphonates of varying purity and composition depending on a number of factors, such as the neutralizing alkali, the pulping process used, the degree of fermentation and even the type and age of the wood used as pulp feedstock [12].

Commercial lignosulphonates used in admixture formulations are predominately calcium or sodium based with sugar contents of 1 to 30%. Typical analyses of two commercially available lignosulphonate water-reducing admixtures are shown in Table 1.2.

The lignosulphonate molecule is a substituted phenyl propane unit

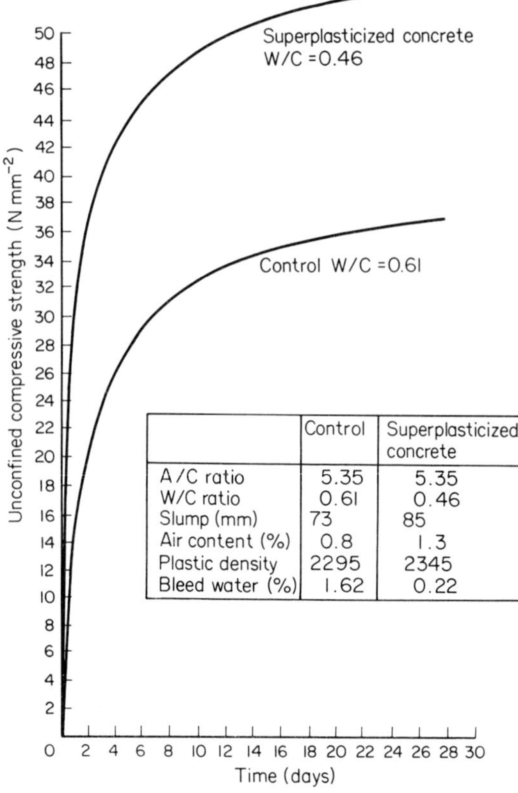

	Control	Superplasticized concrete
A/C ratio	5.35	5.35
W/C ratio	0.61	0.46
Slump (mm)	73	85
Air content (%)	0.8	1.3
Plastic density	2295	2345
Bleed water (%)	1.62	0.22

Fig. 1.4 Relationship between strength and time for a water-reduced concrete containing a superplasticizer.

containing hydroxyl, carboxyl, methoxy and sulphonic acid groups [14-16]. A possible representation is a polymer containing the repeating unit shown in Fig. 1.5. The polymer could typically have an average molecular weight of about 20 to 30 000 with a range varying from a few hundred to 100 000 [13, 17]. The range of molecular weight present in the lignosulphonate is

HO — CH_2 — C(OH) — O — C(H)(SO_3H) — C(O—)—CH_2OH

Fig. 1.5 Repeating unit of a lignosulphonate molecule.

Table 1.1 The formulations of water-reducing admixtures

	Type of water-reducing admixtures			
Normal	Accelerating	Retarding	Air-entraining	Super-plasticizers
Purified lignosulphonate	Lignosulphonate $+CaCl_2$	High sugar lignosulphonate	Impure ligno-sulphonate	Pure ligno-sulphonate
Lignosulphonate + air-detraining agent	Lignosulphonate +triethanol-amine	Hydroxy-carboxylic acid	Lignosulphonate +surfactant	Salt of formaldehyde–naphthalene sulphonate
Hydroxy-carboxylic acid at low dosage	Lignosulphonate +Ca formate	Hydroxylated polymer	Hydroxy-carboxylic acid +surfactant	Salt of formaldehyde–melamine sulphonate
Hydroxylated polymer at low dosage	Hydroxy-carboxylic acid $+CaCl_2$			

dependent on the manner and condition under which refinement is done. Three such methods are used, namely, ultra-filtration, heat treatment at a specified pH and fermentation.

Fig. 1.6 is a typical molecular weight distribution curve for a lignosulphonate obtained by means of gel permeation chromatography (a sophisticated analytical method where molecules are 'sieved' according to their molecular size).

It has been found [18] that the lignosulphonate polymer is not a simple linear flexible coiled 'thread', as found in many high molecular weight materials, but forms spherical microgels of the type shown in Fig. 1.7. Thus the charges are predominately on the outside of the spheroid with the internal carboxyl groups and sulphonate group being non-ionized. Conductivity studies have confirmed that lignosulphonates are only 20 to 30% ionized [17].

Table 1.2 Typical analyses of lignosulphonate-based water-reducing admixtures (after Edmeades)

Type	Spruce wood sulphite lye calcium lignosulphonate (%) [13]	Sodium lignosulphonate (%)
Solid content	54	30
Ash	6.6	—
Sulphated ash	10.9	—
CaO (in ash)	44	0.1
Reducing sugars (as glucose)	5.8	0.9
Total sulphur	3.2	2.6

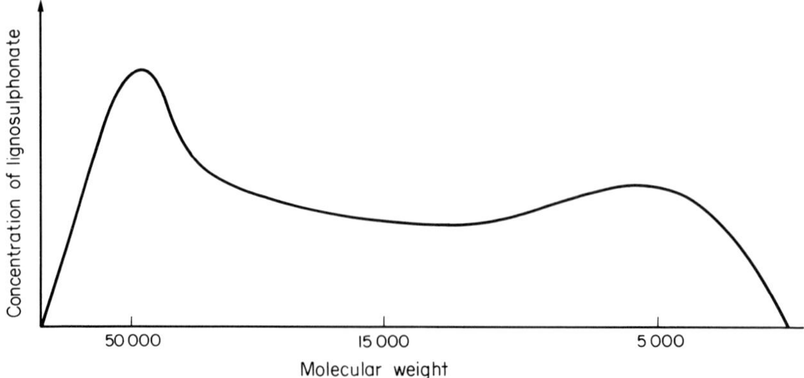

Fig. 1.6 Molecular weight distribution of a typical lignosulphonate.

The *sugars* contained in lignosulphonates vary both in type and concentration, depending on the source, type and degree of refining that has taken place. In the fermentation process the micro-organisms used preferentially consume the hexoses rather than the pentoses so that the residual sugars present in the refined lignosulphonates are mainly pentoses. The types of sugars found are shown in Fig. 1.8 and Table 1.3 gives a breakdown of sugars found in untreated sulphite lye [16] and two commercial water-reducing admixtures [12].

Although several different salts of lignosulphonates are commercially available, the calcium and sodium derivatives are the most widely used in admixture formulation. The sodium salt tends to maintain its solubility at low temperatures, thus avoiding sedimentation in winter conditions. In addition, the sodium salt has a higher degree of ionization in solution [17]

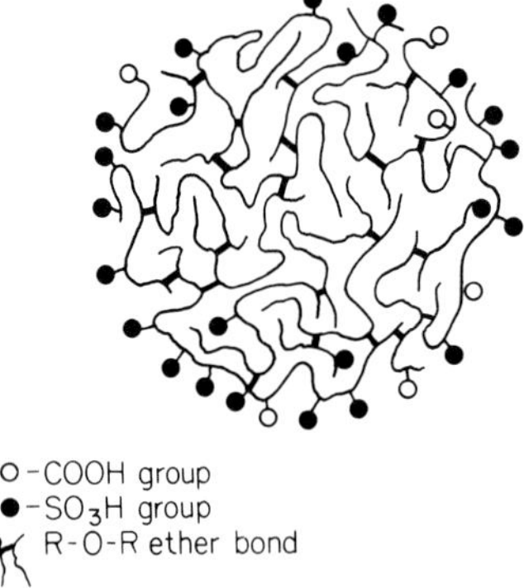

Fig. 1.7 Schematic representation of a lignosulphonate polyelectrolyte microgel unit.

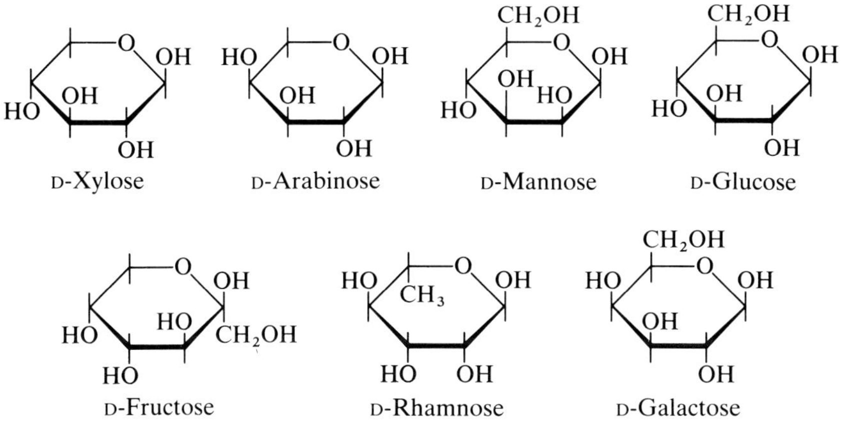

D-Xylose D-Arabinose D-Mannose D-Glucose

D-Fructose D-Rhamnose D-Galactose

Fig. 1.8 Formulae of sugars found in untreated and purified lignosulphonate materials.

Table 1.3 Analysis of sugars in lignosulphonate materials (after Mouton and Joisel)

Material	Sulphitye lye		Fermented admixtures A		B	
Sugar content (%)	30 typical		10.2		5.4	
Composition of sugars (%)						
Pentoses xylose+xylonic acid	15	21	55	70	60	74
arabinose	6		16		14	
Hexoses mannose	48	75			11	26
glucose	15					
fructose	2					
rhamnose	—		16	30	15	
galactose	10		14			
Others	4					

than the calcium salt. This is reflected in the observation that solutions of higher concentrations of the calcium salt are required to obtain the same reduction in water–cement ratio, obtained by using the same dosage of a sodium salt based water-reducing admixture. However, calcium lignosulphonate raw materials are invariably cheaper than sodium lignosulphonates so that the higher concentrations can be offered on an approximately equal cost effectiveness basis.

In the formulation of admixtures from lignosulphonate (see Table 1.1), the following commments are relevant.

(a) Many lignosulphonates and, in particular, the less pure types, entrain a small proportion of air into the concrete. This can be desirable where an air-entraining material is required to enhance durability or cohesion, but is often an unwanted side effect. Thus in the production of normal water-reducing admixtures and superplasticizers, a small quantity of an air-detraining agent is added. The usual material is tributylphosphate at a level of less than 1% of the lignosulphonate, although dibutyl phthalate, water insoluble alcohols, borate esters and silicone derivatives find some application [20].

(b) The lignosulphonate molecule itself and, of course, the sugars present in the lignosulphonate materials do have a retarding influence on the hydration of the cement. In the case of the higher sugar content materials, this is utilized to produce the retarding water-reducing admixtures which allow longer transport or placing times. However, for normal water-reducing admixtures, this is an undesirable effect and, therefore, additions of triethanolamine are occasionally made at a level of about 15% of the lignosulphonate content of the admixture [19]. At this level of addition the triethanolamine acts as an accelerator and compensates for the retarding influence of the lignosulphonate and its impurities. This has been shown to have certain deleterious effects on some properties of the resultant concrete (see Sections 1.5.2 and 4.5.2).

(c) The accelerating water-reducing admixtures are simple blends of either calcium chloride or calcium formate with a lignosulphonate. In both cases it is not possible to obtain a completely sediment-free solution and agitation of store tanks may be necessary. Typically, a mixture of approximately 33% calcium chloride and 4% calcium lignosulphonate by weight in water would be used.

(d) Air-entraining water-reducing admixtures containing lignosulphonates can be based on impure lignosulphonate raw materials, as stated earlier, where only 2 to 3% additional air is required. However, this air may not be of the amount, type, and stability required (see Sections 2.3.2 and 2.4.2), therefore additions of surfactants are made. Several different types can be used but in the majority of cases they are based on alkyl–aryl sulphonates (e.g. sodium dodecyl benzene sulphonate) or fatty acid soaps (e.g. the sodium salt of tall oil fatty acids). Additions of these types will allow incorporation of sufficient stable air of the correct bubble size to meet durability requirements under freeze–thaw conditions.

1.2.2 Hydroxycarboxylic acids

As the name implies, these are organic chemicals which have both hydroxyl and carboxyl groups in their molecules. Generally, the sodium salt is used, although occasionally the materials are found as salts of ammonia or triethanolamine. They are produced from pure raw materials

	Material Citric acid [24, 26]	Tartaric acid [24, 26]	Mucic acid [11, 15, 26]
Functionality OH groups	1	2	4
COOH groups	3	2	2
Molecular weight	192	150	210
Formula	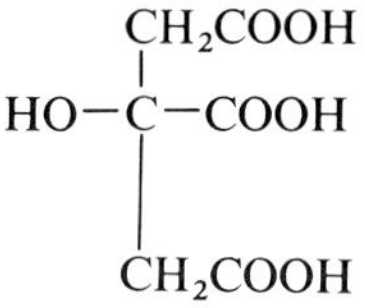 CH_2COOH | HO—C—COOH | CH_2COOH	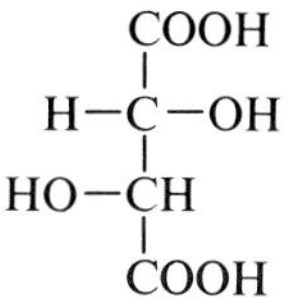 COOH | H—C—OH | HO—CH | COOH	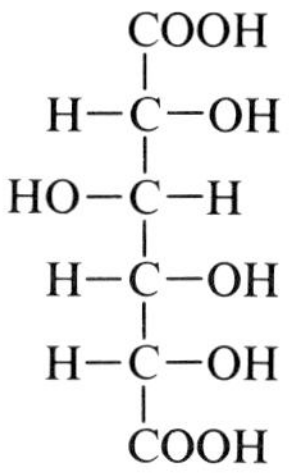 COOH | H—C—OH | HO—C—H | H—C—OH | H—C—OH | COOH

	Gluconic acid [11, 21, 22, 27]	Salicylic acid [25]	Heptonic acid [23]	Malic acid [26]
Functionality OH groups	5	1	6	1
COOH groups	1	1	1	2
Molecular weight	196	138	230	134
Formula	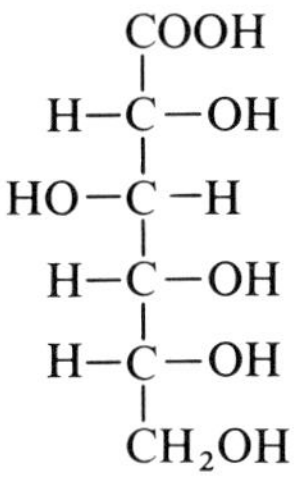 COOH | H—C—OH | HO—C—H | H—C—OH | H—C—OH | CH_2OH	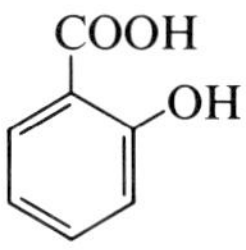 COOH (benzene ring) OH	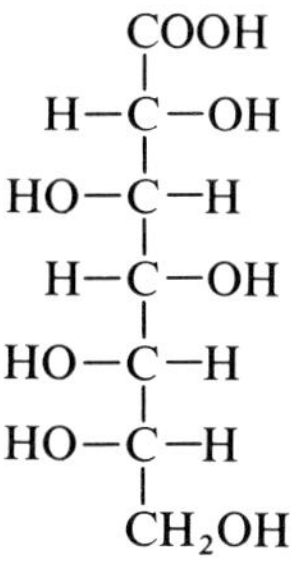 COOH | H—C—OH | HO—C—H | H—C—OH | HO—C—H | HO—C—H | CH_2OH	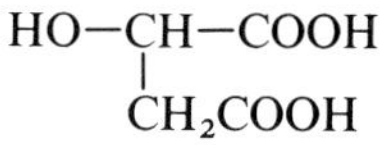 HO—CH—COOH | CH_2COOH

Fig. 1.9 Hydroxycarboxylic acids used in admixtures.

feedstocks, by either chemical or biochemical means and, therefore, are of high and consistent purity. Indeed, the primary use of the materials is often in foodstuffs or pharmaceuticals.

In the form of soduim salts all are very soluble and have low freezing points, so that solidification in winter conditions is unlikely. Fig. 1.9 shows the types and formulae of materials which find application in the formulation of this type of water-reducing admixture.

Normally, approximately 30% solutions of the salts would be used with additions of other chemical types, depending on the proposed function in concrete. Thus the salts may be present alone to produce normal water-reducing admixtures at low dosages and retarding water-reducing admixtures at higher dosages. Small amounts can be blended with calcium chloride to produce accelerating water-reducing admixtures which are almost colourless, sediment-free solutions and, in a similar manner to lignosulphonates (see earlier), air-entraining agents can be added to form the air-entraining water-reducing admixtures which may or may not be retarding, depending on the amount of hydroxycarboxylic acid salt present in the formulation.

The hydroxycarboxylic acid materials are not used in the formulations of superplasticizers because of the considerable retarding influence that would be experienced at the high dosages necessary to produce the required workability.

1.2.3 Hydroxylated polymers

The hydroxylated polymers are derived from naturally occurring polysaccharides, such as corn starch by partial hydrolysis to form lower molecular weight polymers containing from 3 to 25 glycoside units [28].

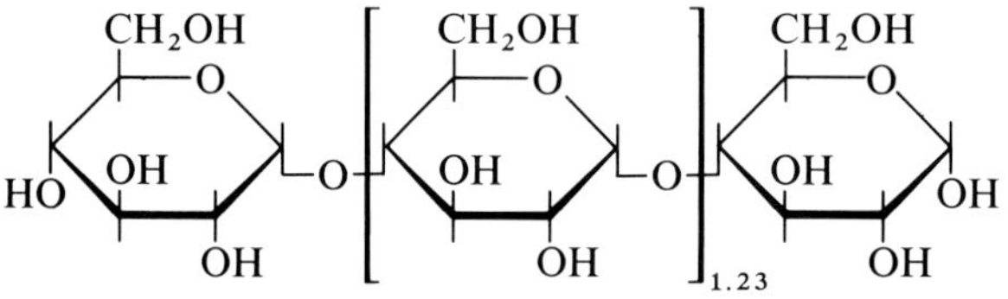

Fig. 1.10

Unlike the monosaccharide glucose, these materials are stable under the alkaline conditions of a cement-containing composition and behave as efficient water-reducing agents. They do impart a retardation to the concrete in which they are incorporated, which can be overcome by the addition of small quantities of calcium chloride or triethanolamine [28].

1.2.4 Salts of naphthalene formaldehyde sulphonic acids

This raw material was one of the first referred to in the literature as a water-reducing agent [1], yet only in recent years has it found major application in admixture formulations [29–35].

The material is produced from naphthalene by oleum or sulphur trioxide sulphonation. Subsequent reaction with formaldehyde leads to polymerization and the sulphonic acid is normally neutralized with sodium hydroxide [36].

The value of *n* (repeating unit) is typically low and predominantly the dimer is formed. The quantity of sodium sulphate by-product formed by

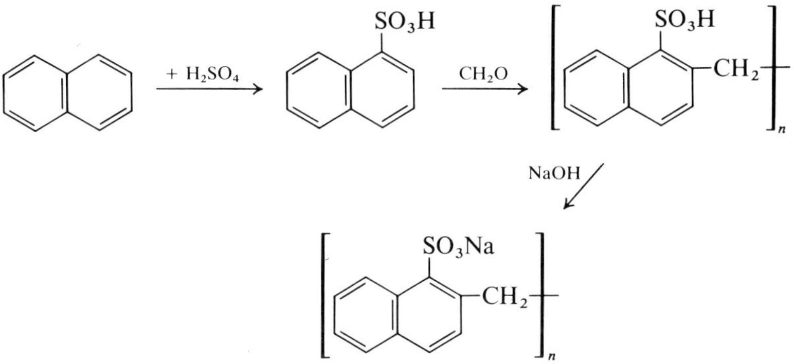

Fig. 1.11

neutralization of excess sulphonating reagent will vary on the process used, but can be reduced by a subsequent precipitation process using lime [37].

The commercially available materials of low polymerization tend to reduce the surface tension of the aqueous phase in the concrete which can lead to air entrapment. In view of this, air-detraining agents, such as tributyl phosphate and dibutyl phthalate are added. Materials based on higher molecular weights are said to be more effective and as they do not affect the surface tension, will not cause air entrapment [34]. In such materials, the value of n would be about 10.

This raw material is used mainly to produce superplasticizers because it is possible to add large amounts to cement-containing compositions, so that considerable increases in workability, or reductions in water–cement ratio can be obtained without the undesirable side effects of severe retardation or air entrainment. Commercial products vary from 25 to 45% w/w solids contents and addition levels required to produce concrete of almost self-compacting properties would be 1.0 to 3.0% by weight of the cement content.

1.2.5 Salts of melamine formaldehyde sulphonates

This type of chemical product was originally developed [38] in the 1950s for applications in a variety of industries, but it was not until about ten years later that the possibility for application in the concrete field was recognized [39].

It is prepared by normal resinification techniques according to the process given [40] in Fig. 1.12. The normal production procedure results in a product conforming to the characteristics given in Table 1.4 [41].

The length of polymerization time will be reflected in the average molecular weight of the product, but the most useful product is believed to have an average molecular weight of about 30 000.

Although the material is used almost entirely as the sole ingredient in superplasticizers, some interest has been shown in blending with hydroxycarboxylic acids [42] or lignosulphonates. In its role as a superplasticizer it conforms to the requirements of no excessive retardation or air entrainment at high dosage levels.

The five categories of major ingredients, discussed above for the formulation of water-reducing admixtures, account for the majority of commercially available products, but there may be limited use of insitol [43], polyacrylamide [44], polyacrylic acids [45], and polyglycerol [46].

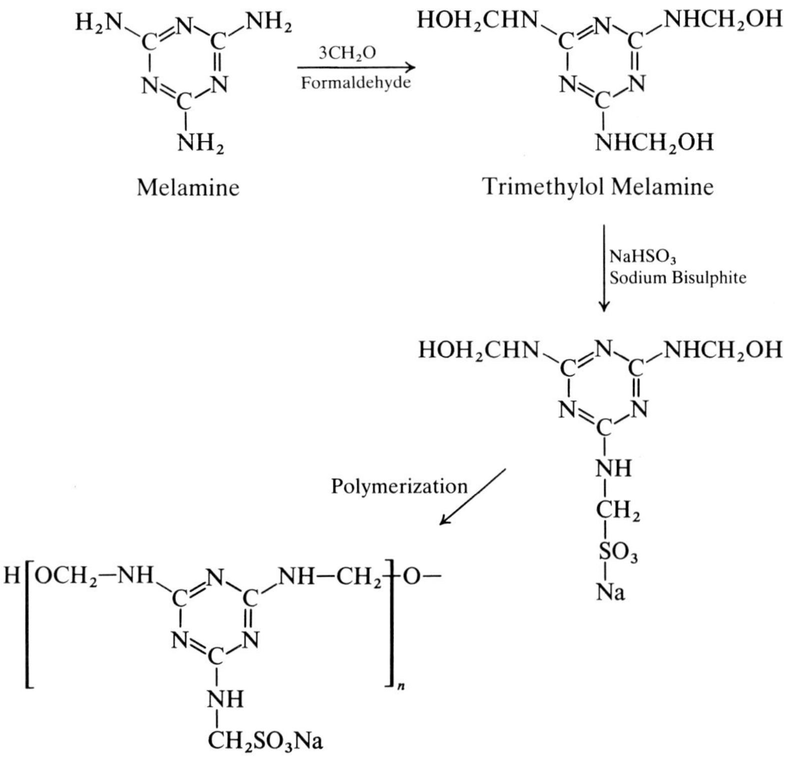

Melamine $\xrightarrow[\text{Formaldehyde}]{3CH_2O}$ Trimethylol Melamine $\xrightarrow[\text{Sodium Bisulphite}]{NaHSO_3}$ (sulphonated methylol melamine, $-NH-CH_2-SO_3Na$) $\xrightarrow{\text{Polymerization}}$ $H[OCH_2-NH-(\text{triazine ring, } NH-CH_2SO_3Na)-NH-CH_2]_n-O-$

Fig. 1.12

Table 1.4 Typical characteristics of an anionic melamine formaldehyde resin (after Davis)

Solids content (%) w/w	20
Viscosity (cP) at 20° C	10
pH	8.5
Specific gravity at 20° C	1.12
Appearance	clear water white solution

1.3 The effects of water-reducing admixtures on the water–cement system

In order to understand more fully the effect that water-reducing admixtures have on the plastic properties of fresh concrete, and to gain an insight into the mechanism of action of this category of materials, it is useful to study the effect on the water–cement system. The topic can be considered from the following points of view: (a) rheological and initial surface effects and (b) effects on soluble and insoluble hydration products and kinetics of hydration.

1.3.1 Rheological considerations

It is known that some of the properties of fresh concrete can be considered in terms of the rheological properties of the cement paste contained in the concrete. Thus a high water–cement ratio concrete will contain a paste content which is more fluid than that of a low water–cement ratio concrete.

The 'fluidity' of the cement paste can be measured in rheological terms by the torque transmitted to a stationary 'bob' inside a revolving outer cylinder placed in a water–cement system as shown in Fig. 1.13. The shear stress measured at the stationary bob is plotted against the rate of applied shear when, for pastes of varying water–cement ratios the results shown in Fig.

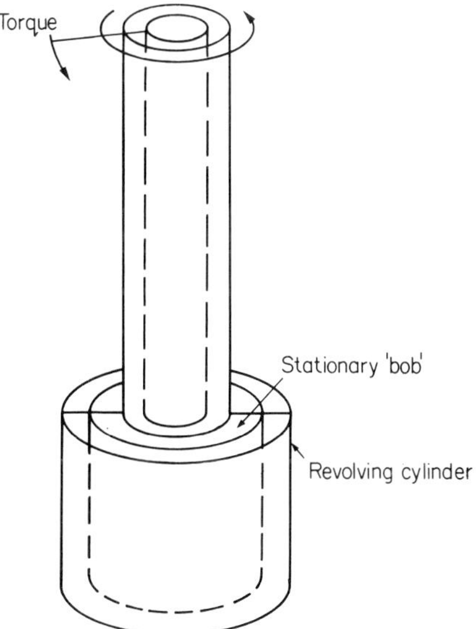

Fig. 1.13 A concentric cylinder viscometer.

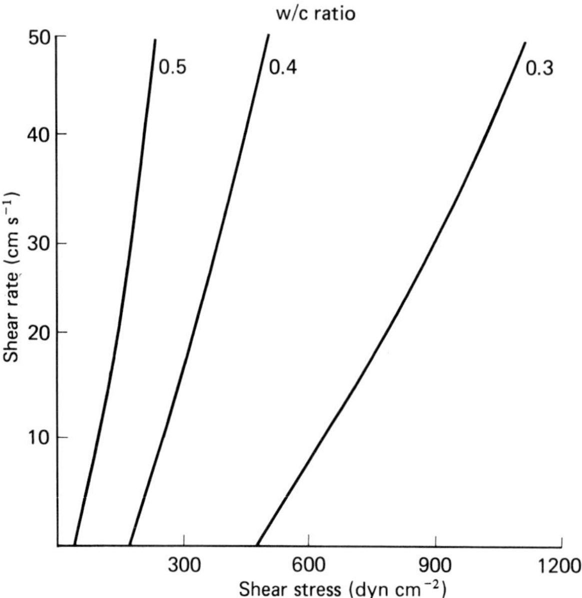

Fig. 1.14 Shear stress/shear rate relationships for cement pastes at various water–cement ratios.

1.14 are obtained for readings taken of the shear stress as the shearing rate is increased (the up curve).

Systems which give linear shear stress–shear rate relationships with an intercept on the shear stress axis are known as exhibiting plastic flow, and the intercept value is known as the 'yield stress'. The viscosity of the system is given by the slope of the line over its linear portion.

The properties of an overall concrete system in the plastic state will be a function of many parameters such as aggregate types and shapes, cement contents and characteristics, etc., but it is useful to isolate the effect of the paste rheology where it can be stated that:

The *consistency* or *fluidity* of the concrete will be a function of the viscosity of the cement paste.

The *cohesion* of the concrete will be a function of the yield stress of the cement paste.

Turning now to the effect that water-reducing admixtures have on the rheology of cement pastes, it can be seen from Fig. 1.15 that the addition of these types of materials does not appear to alter the shape of the shear stress–shear rate relationship but merely moves it to a lower level (the lignosulphonate material has had tributylphosphate added to it so that the effect of air-entrainment is eliminated).

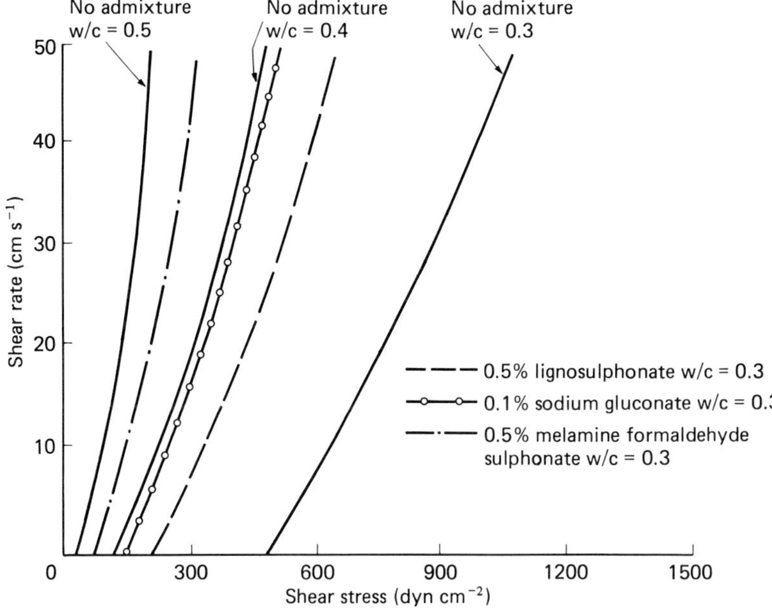

Fig. 1.15 Shear stress/shear rate relationships for cement pastes containing various water-reducing agents.

In view of the relationship shown in Fig. 1.15 it is useful to use paste viscosities as a means of assessing and studying water-reducing admixtures and Figs 1.16 and 1.17 show, respectively, the effect of varying addition levels of three water-reducing admixtures on the paste viscosity, and the effect of the three materials at typical dosages at various water–cement ratios. These data would indicate that the water reductions possible using the three types of water-reducing agents are different and depend on the water–cement ratio of the system. Typical results are given in Table 1.5 and these values are somewhat higher than those obtained in concrete mixes emphasizing the importance of other mix parameters. The lignosulphonates

Table 1.5 Typical water reductions of cement paste at normal addition levels of admixture to maintain paste viscosity

Admixture at normal dosage rate	% Average water reduction over w/c 0.3 to 0.5
Sodium lignosulphonate	10
Sodium gluconate	16
Sodium naphthalene sulphonate formaldehyde polymer	26

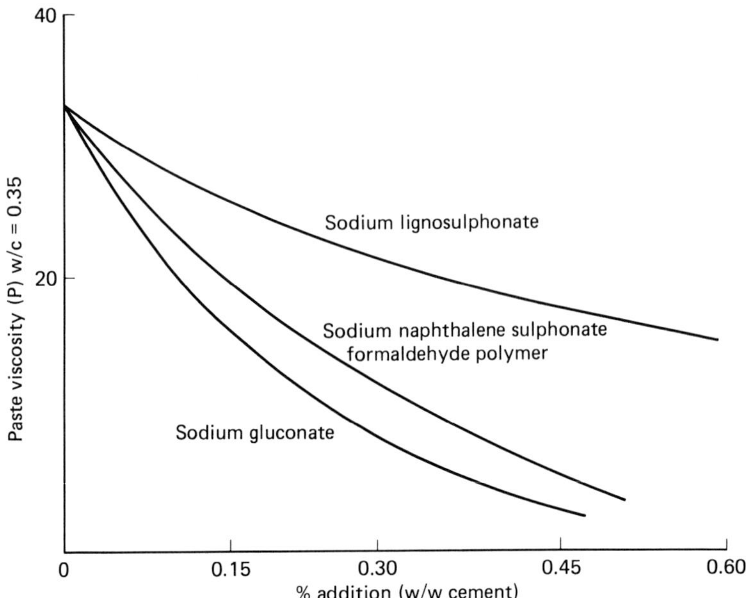

Fig. 1.16 The effect of water-reducing admixtures on paste viscosity at different addition levels.

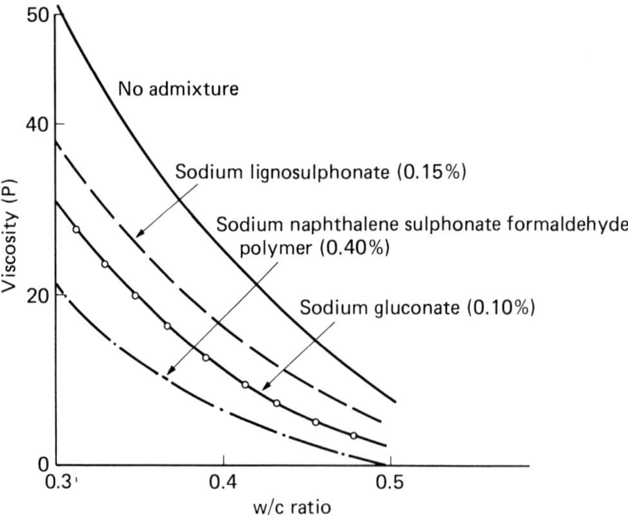

Fig. 1.17 The effect of water-reducing admixtures on paste viscosity at various water–cement ratios.

also entrain some air into concrete which could increase the water reduction obtained.

1.3.2 Initial surface effects

The rheological characteristics of cement pastes are related to the nature of the attractive and repulsive forces which exist between cement and cement hydration product particles and can be categorized as follows:

(a) Van der Waals forces of attraction which are large in magnitude but only at interparticle distances of up to 50 to 70 Å.
(b) Electrical repulsion due to the cationic nature of the free valences at the surface of the cement particles due to the Ca, Al and Si atoms [47]. This repulsion is smaller in magnitude than the van der Waals forces but, because of an associated ion and water molecule 'sheath', probably exists to a significant level at interparticle distances of up to 150 Å

Thus in normal cement pastes where particles come into close contact with each other there is a tendency for cement pastes to form large 'flocs' due to the attractive van der Waals forces holding particles together.

In order to understand how the addition of small amounts of organic materials can reduce this interparticle attraction to an extent that considerable quantities of water can be removed from the system whilst maintaining the same rheological characteristics, it is necessary to consider a large amount of experimental data:

(1) The addition of a water-reducing agent to the aqueous phase is followed by a rapid reduction in the quantity of admixture in solution and an increase in the amount adsorbed at the cement particle/water interface [28, 48–54]. This effect is illustrated by an adsorption isotherm where various addition levels of admixtures are added to the water–cement systems and shaken for a period of time when the amount left in solution is estimated. The amount of material on the surface is calculated by difference. A typical curve for calcium lignosulphonate on ordinary Portland cement is given in Fig. 1.18 [48].

(2) Water-reducing admixtures are not adsorbed equally by the various anhydrous and hydrated cement constituents and in studies with calcium lignosulphonate, the approximate maximum adsorption figures shown in Table 1.6 have been obtained [53, 54]. In addition, adsorption isotherms have been studied at various ages of C_3A hydration [51] and it has been shown that it is the initial hydration products (less than 15 min) which have the highest adsorptive capacity for calcium lignosulphonate suggesting that it may be the precursor to the hexagonal phases which is responsible for the initial high adsorption, i.e. C_4AH_{19}. Fig. 1.19 indicates this effect. Data for

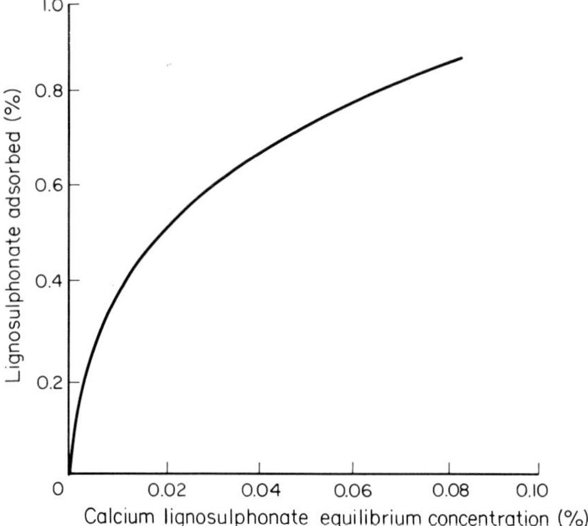

Fig. 1.18 Adsorption isotherm of calcium lignosulphonate on ordinary Portland cement (Ernsberger).

Table 1.6 Adsorption of calcium lignosulphate by cement consistuents (after Feldman and Ramachandran)

Cement constituent*			% Adsorption
C_3A (anhydrous)			0
C_3A hydrates–	C_2AH_8	hexagonal	2.2
	C_4AH_{13}	hexagonal	
	C_3AH_6	cubic	0
CH			12
C_3S			0
C_3S hydrates			7

* The usual abbreviations are used here and throughout this book: $C = CaO$, $A = Al_2O_3$, $S = SiO_2$, $H = H_2O$.

salicylic acid (a hydroxycarboxylic acid) yield similar information although adsorption levels are usually lower.

(3) The various types of water-reducing admixtures possess different but characteristic adsorption isotherms which qualitatively reflect their effect on cement hydration kinetics, as shown in Fig. 1.20.

(4) In the absence of knowledge of the surface area of cement hydrates available for adsorption at the time of addition, it is difficult to estimate how many layers of water-reducing admixture molecules are adsorbed, but attempts have been made [55] indicating that over 100 layers may be formed with calcium lignosulphonate and salicylic acid at normal levels of addition. However, these calculations were based on specific surface areas of 0.3 to

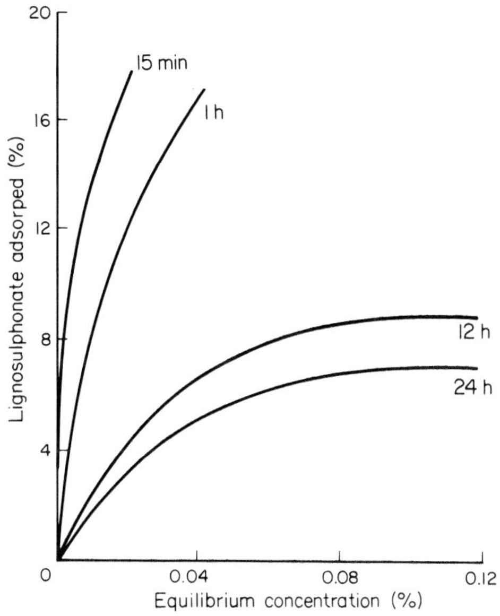

Fig. 1.19 Effect of hydration time on the adsorption of calcium lignosulphonate on C_3A hydrates (Rossington).

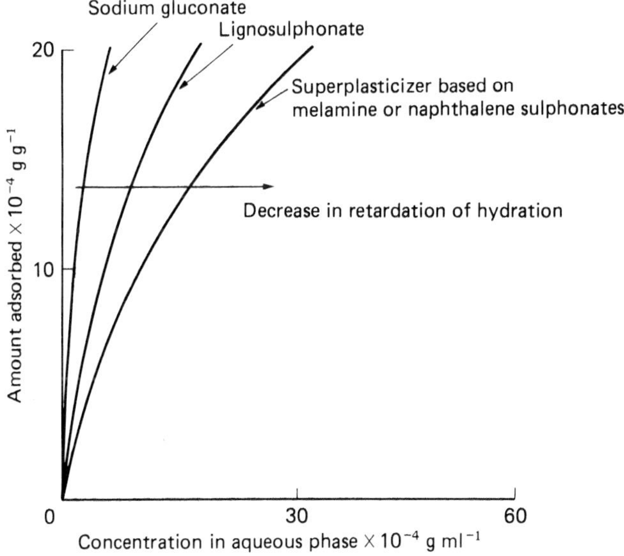

Fig. 1.20 Adsorption isotherms for various water-reducing admixtures.

Table 1.7 Surface areas of cement constituent hydrates (Feldman and Ramachandran)

Phase	Surface area ($m^2\ g^{-1}$)	Adsorbed gas
C_3A hydrate (hexagonal)	11–14	N_2
C_3S hydrate	136	H_2O
C_3S hydrate	70	N_2
Calcium hydroxide	16	—

Table 1.8 Adsorption data for calcium lignosulphonate and sodium gluconate

Material	Molecular area (Å^2)	Surface covered by 1 mg material (m^2)	Addition level for min. paste viscosity (%)	Amount adsorbed (from adsorption isotherm) (%)	Area for monolayer completion at min. paste viscosity ($m^2 g^{-1}$)
Calcium lignosulphonate	250 000	0.6	1.0	0.6	3.6
Sodium gluconate	50	1.5	0.5	0.45	6.8

1.0 $m^2\ g^{-1}$ whereas other studies [28, 53, 54] have indicated considerably higher surface areas as shown in Table 1.7. The earlier C_3A hydrates would be expected to have an even higher surface area, although all the surface available to nitrogen molecules would not be available to the larger admixture molecules. For two materials, calcium lignosulphonate and

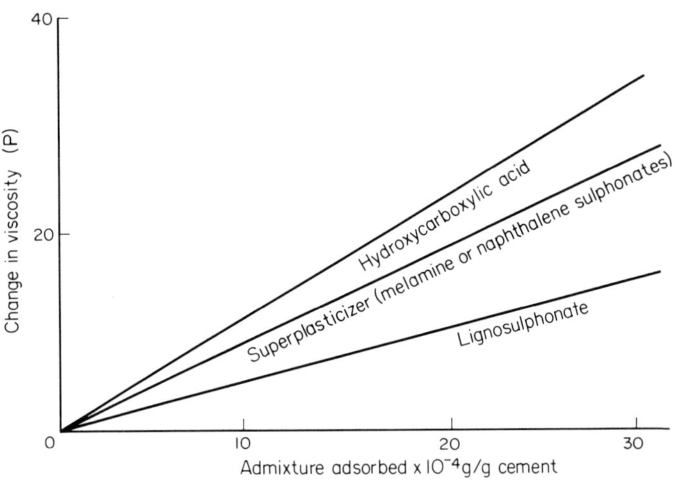

Fig. 1.21 Reduction in cement paste viscosity by various water-reducing admixtures as a function of amount adsorbed.

sodium gluconate, it is possible to compile the data shown in Table 1.8 from which it can be seen that a surface area available to the admixtures of about 3 to 7 $m^2\ g^{-1}$ would be required for formation of a monolayer of admixture molecules. The values given in Table 1.7 suggest that even after a few minutes hydration, this level of available surface may be present. At levels much higher than the addition required for the minimum paste viscosity, it is likely that multilayers are formed, possibly of insoluble aluminium salts.

(5) In the case of those materials known to possess an adsorptive capacity towards cement constituents and the ability to reduce the paste viscosity, there is a linear correlation between the amount adsorbed on the surface and the reduction in paste viscosity at addition levels in the region of those normally used. This is shown in Fig. 1.21. These data can be used to obtain a viscosity reducing index for various materials by measuring the slope of the line. Some values are shown in Table 1.9 although it should be emphasized that this technique is only used by the author as a means of comparing series of experimental products with known standards since different values are obtained for different batches of cement.

Table 1.9 Viscosity reduction of cement pastes by various water-reducing admixtures

Material	Viscosity reduction index (P $m^{-1}\ g^{-1}$) (w/c ratio = 0.30)
Calcium lignosulphonate	5
Melamine formaldehyde sulphonate Naphthalene formaldehyde sulphonate Polysaccharide	9
Sodium gluconate	12

(6) The nature of the bond between the molecules of the water-reducing admixture and the surface of the cement constituent hydrates has been described as 'ionic group outwards' in many references [33, 48, 56,] mainly based on early work [48] showing migration of cement particles under the influence of an electric current when lignosulphonate molecules are adsorbed on the surface. The lignosulphonates may, however, be a special case because of their spherical shape; it would be possible for the molecules to bond 'ionic group inwards' and yet still possess outward pointing ionized and non-ionized groups on the other side of the spherical molecule. Other relevant data are summarized below.

(a) All classes of water-reducing admixtures have been found to be irreversibly adsorbed on to hydrating Portland cement including sulphonated naphthalene formaldehyde products which possess only sulphonate functional groups. This is shown experimentally by the addition of further quantities of solvent during the determination of the

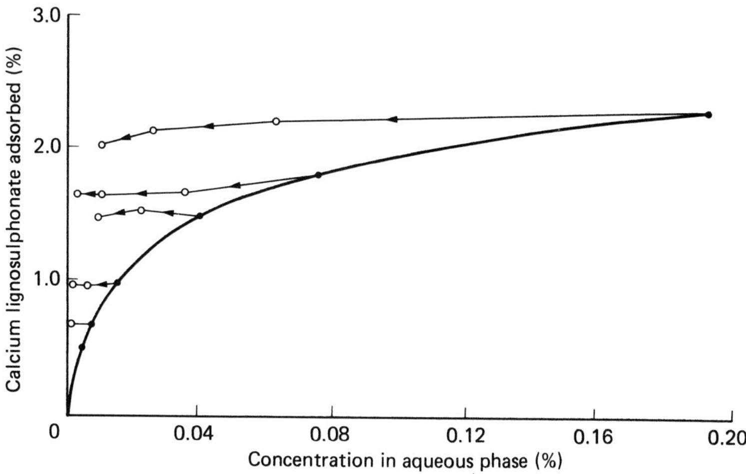

Fig. 1.22 Adsorption–desorption isotherms of calcium lignosulphonate on C_3A hydrate (hexagonal phase) (Ramachandran).

adsorption isotherms; in true physical adsorption the original isotherms should be followed, but as shown for calcium lignosulphonate on the C_3A hexagonal hydrates (Fig. 1.22) there is almost no tendency for desorption [53]. The adsorption has also been shown to be irreversible on hydrated C_3S phase for calcium lignosulphonate materials [54].

(b) The amount of calcium lignosulphonate adsorbed on to hydrating cement is almost independent of initial water–cement ratio within the range 0.4 to 1.5 [49].

(c) In model studies with an hydroxycarboxylic acid (salicylic acid), a co-ordination complex has been isolated from the reaction with the C_3A hydrates [57]:

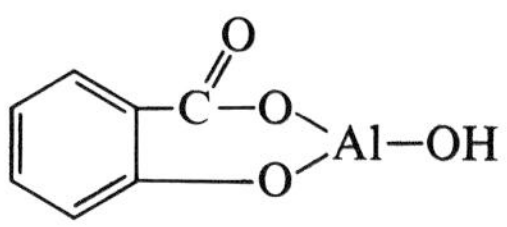

In addition, infra-red evidence exists for the formation of a bond between the sulphonate group and the C_3A phase hydrate for the sulphonated melamine formaldehyde superplasticizers [58], and differential thermal analysis has indicated the formation of a complex between C_3S hydrates and calcium lignosulphonates [54].

(d) The value and sign of the surface potential of Portland cement in the presence of varying amounts of ammonium lignosulphonates are shown in Table 1.10 [59]. Comparing these data with a typical adsorption isotherm would support a progressive surface coverage up to a level of

Table 1.10 Reduction of surface potential by ammonium lignosulphonate (after Zhuravhev)

% Ammonium lignosulphonate on cement weight	0	0.01	0.05	0.1	0.25	0.8	1.0
Surface potential (mV)	+7.2	+6.0	+5.8	+4.1	+3.5	+3.3	+1.1

0.25 to 0.5% of lignosulphonate where an imperfect monomolecular layer would be formed.

It can, therefore, be concluded with some confidence that water-reducing admixtures chemically bond through their carboxyl, sulphonate, and/or hydroxyl groups to surface calcium or aluminium ions as illustrated in Fig. 1.23.

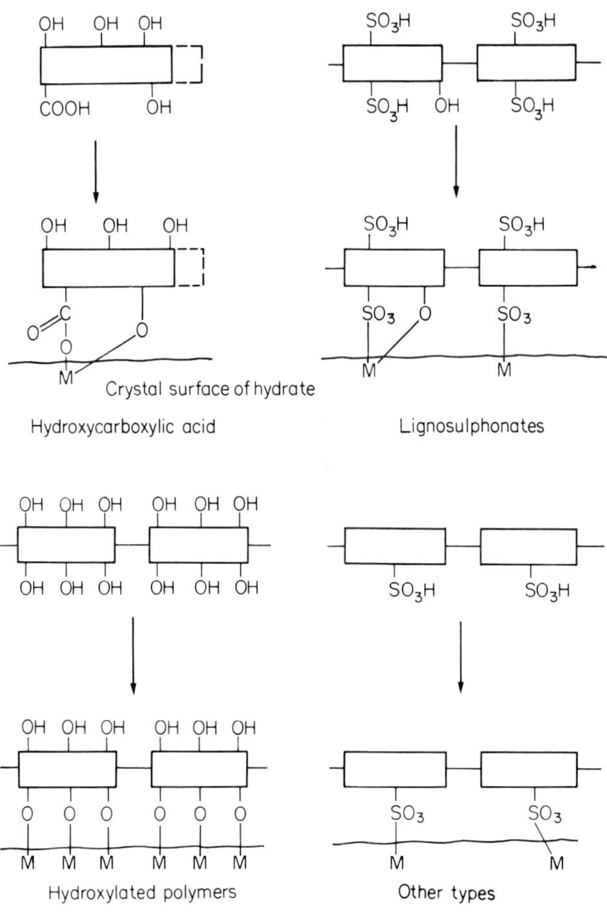

Fig. 1.23 Diagrammatic scheme of adsorption.

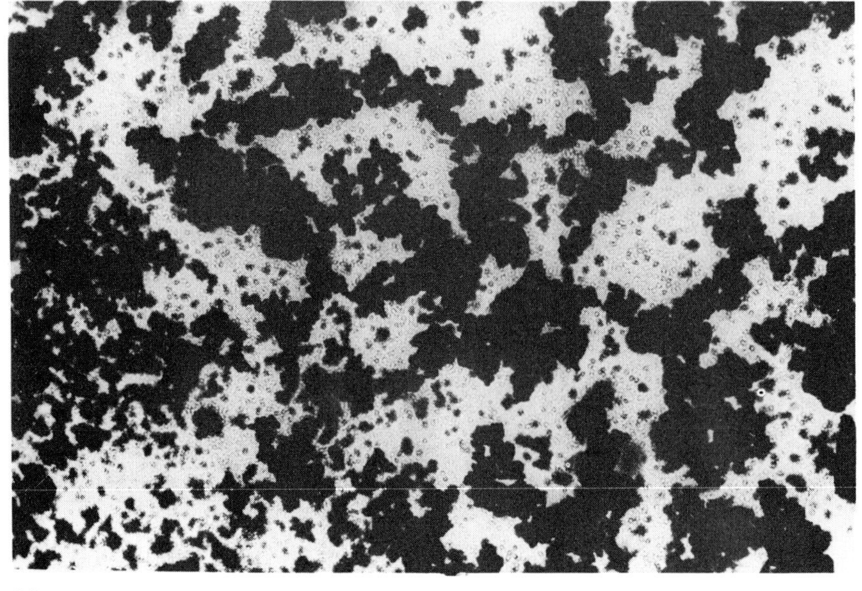

(a)

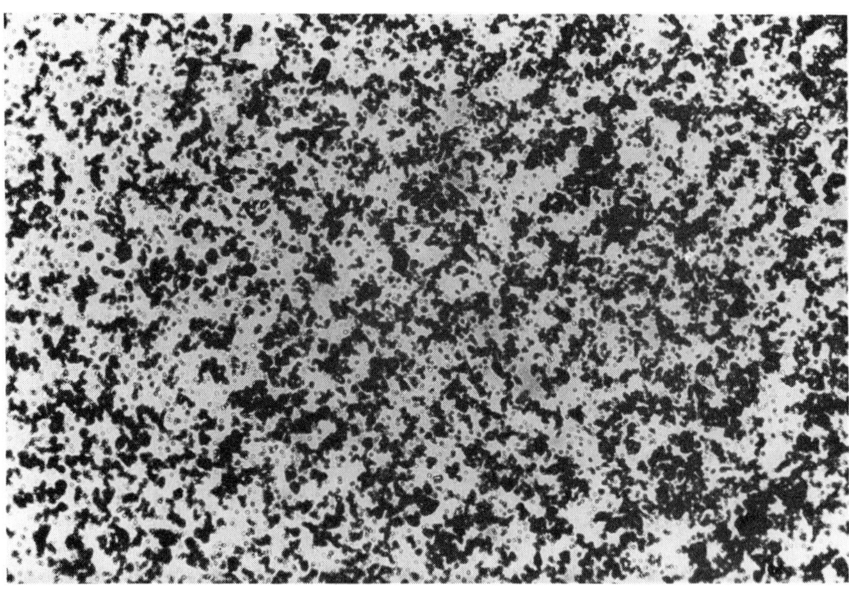

(b)

Fig. 1.24 Dispersion of cement particles by a water-reducing admixture, (a) before addition, (b) after addition.

In the case of lignosulphonates, the intermediate adsorption isotherms indicate that there may be some reaction involving hydroxyl groups and the nitrogen atoms of the sulphonated melamine formaldehyde may be capable of either co-ordination complex formation or strong physical adsorption.

(7) It seems likely that the various types of water-reducing admixture molecules, either by virtue of their high hydroxyl content (hydroxylated polymer and hydroxycarboxylic acid), electrical double layer of microgels (lignosulphonates) [18], or high electron density of aromatic character systems (sulphonated naphthalene or melamine formaldehyde materials), will be highly solvated and, therefore, surrounded by a 'sheath' of associated water molecules.

(8) The addition of a water-reducing admixture to a cement suspension can be shown to disperse the agglomerates of cement particles into smaller particles [34, 48, 53, 60] and can be seen clearly in photomicrographs as shown in Fig. 1.24. Maximum dispersion occurs at a level of 0.3 to 0.5% by weight of calcium lignosulphonate [48, 49] which would indicate the presence at the surface of about 0.2 to 0.4% calcium lignosulphonate. The separation of particles results in an increase in the surface area of the system by 30 to 40% [48, 53] which may explain the more rapid rate of cement hydration after the initial retardation period.

1.3.3 Effects on the products and kinetics of hydration

Aqueous phase hydration products

During the hydration of Portland cement the concentration and nature of dissolved ions in the aqueous phase is constantly altering particularly at early hydration times. The major ionic species present are the alkali metals sodium and potassium, the alkaline earth metal calcium, and the anions sulphate and hydroxyl. Fig 1.25 illustrates the level of these anionic species present over a period of time. Also shown are the changes in these anionic concentrations which would occur where an addition of a fairly high dosage of lignosulphonate is made or a normal dosage of sodium gluconate type water-reducing admixture [49, 52, 61–63]. The following observation can be made:

(a) The calcium ion concentration of the solution phase is slightly increased in the early stages of hydration, but subsequently the concentration approaches that of a system containing no admixture. It has been found that the greater the period of set retardation, the larger is the difference in calcium concentration, and it takes longer to reach the same level as a control system. The retarded hydration of the C_3S phase would expect to be reflected in a lower concentration, but it is believed

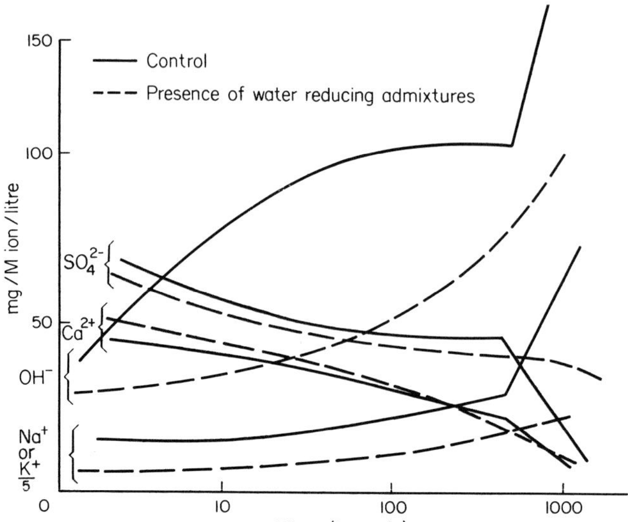

Fig. 1.25 The level of various anionic species present in the aqueous phase of hydrating cement pastes.

that the observed increase is due to a delay in a drop of the calcium and sulphate ion concentrations from the supersaturated to the saturated level. As can be seen this drop occurs very early in the control water–cement system.

(b) The sulphate ion concentration normally diminishes rapidly, whereas in the presence of a water-reducing admixture, the high concentration is maintained due to the reasons given in (a) which, in turn, can be related to a delay in the reaction of the gypsum by the C_3A phase to form ettringite. Indeed, if the cement constituents are given a pre-hydration period prior to the addition of the admixture, results intermediate between those shown in Fig. 1.25 are obtained because part of the $C_3A + CaSO_4$ reaction has been permitted to proceed [62].

(c) The hydroxyl ion concentration is initially reduced due to the retardation of the C_3S hydration to form $Ca(OH)_2$ and the sudden increase in hydroxyl ion concentration is smoothed out probably due to the gradual breakdown of the inhibiting admixture monolayer to give a faster diffusion of hydration products.

(d) The alkali metal ions Na^+ and K^+ behave in a similar manner to the hydroxyl ion concentration, again due to the delay in C_3S hydration since the majority of the soluble alkaline metal ions originates from the C_3S crystalline phase.

These observations on the aqueous phase are consistent with the concept of the adsorption of water-reducing admixtures on to the initial hydrates of C_3A and C_3S with a corresponding modification of the normal process of reaction of C_3A with gypsum to form ettringite and a delay in hydration of the C_3S phase.

Effect on solid hydration products

The hydration process of Portland cement is extremely complex but can be largely considered in terms of the hydration of the major phases present and reactions with calcium sulphate. A number of studies have been made of the various phases and it is generally considered that the behaviour of a typical Portland cement is largely that of the sum of its components [64]. The changes on addition of water-reducing admixtures can be considered in terms of the chemistry, morphology and reaction kinetics. Fig. 1.26 shows a typified heat evolution carried out under isothermal conditions using a sample of a normal Portland cement containing C_3A, C_2S, C_3S and gypsum [65].

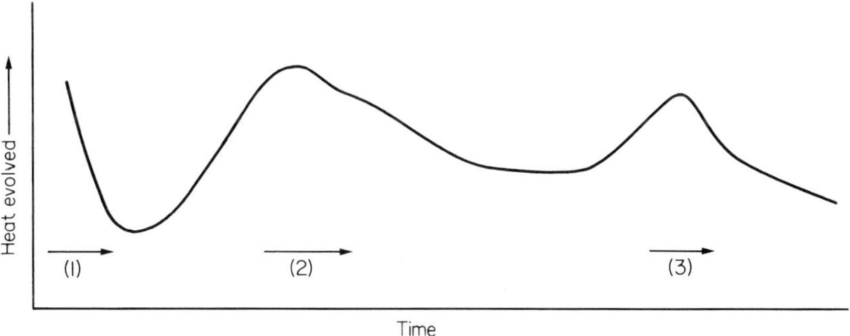

Fig. 1.26 Isothermal calorimetric curve of hydrating Portland cement.

The chemistry associated with peaks 1, 2 and 3 can be summarized as follows:

Peak 1

(a) $C_3A + H_2O = C_4AH_{19}$ (quickly stopped by formation of a protective coating of ettringite).

(b) $C_3A + CaSO_4 = C_3A.3CaSO_4.32H_2O$.

(c) $C_3S + H_2O =$ C–S–H initial C–S hydrate layer about 100 Å thick and low in Ca.

(d) Heat of solution of free alkali.

The C_3A phase is subjected to competitive reaction between water and gypsum to form a small amount of the initial C_3A hydrates covered by a protective coating of ettringite, and also a proportion of the C_3S hydrates to form a calcium silicate hydrate of low calcium content in the region of 100 Å thick.

Peak 2

C_2S, C_3S	$+ H_2O$	C—S—H (gel) approx $C_3S_2H_{25}$	plates C : S = 0.75–1.0 : 1.0 800 Å × 20 Å in twos or threes
		+	needles C : S = 2 : 1 (probably from C_3S)
		$Ca(OH)_2$ hexagonal prisms	crumpled foils C : S = variable 10 to 20 Å thick

During this phase the initial self-retarding calcium silicate hydrate layer appears to break down allowing further hydration products from the C_3S and C_2S phases to be formed. These appear to be produced in three morphological forms, plate-like, needles and crumpled foils. During this reaction considerable quantities of hexagonal prisms of precipitated lime are also formed.

Peak 3

In this final peak, the remaining tricalcium aluminate reacts with both gypsum and water to form ettringite and subsequently the tricalcium aluminate monosulphate. C_3A hydrates formed are of varying composition but eventually form the stable cubic phase C_3AH_6 in solid solution with the tricalcium aluminate monosulphate.

There has been a considerable study of the effect of various water-reducing admixtures on the pure phases and also on ordinary Portland cement. The following points summarize the general observations.

(a) In the presence of a sodium lignosulphonate water-reducing admixture having the following composition: 57.6% lignosulphonic acid, 11.6% reducing sugars, 18.5% ash, the effect on the heat evolution curve under isothermal conditions is shown in Fig. 1.27. These results [66, 67] indicate that the second peak, i.e. the hydration of the C_2S and C_3S phases, is retarded whilst that of the third peak is accelerated. Some results have been obtained for sodium gluconate [68] at least up to the end of the second peak. It seems, therefore, that the C_2S and C_3S phases are retarded and that the presence of a water-reducing admixture interferes with the ettringite reaction particularly in the conversion of the ettringite to the monosulphate.

(b) The point in (a) is borne out by the analysis of mixtures of C_3A and

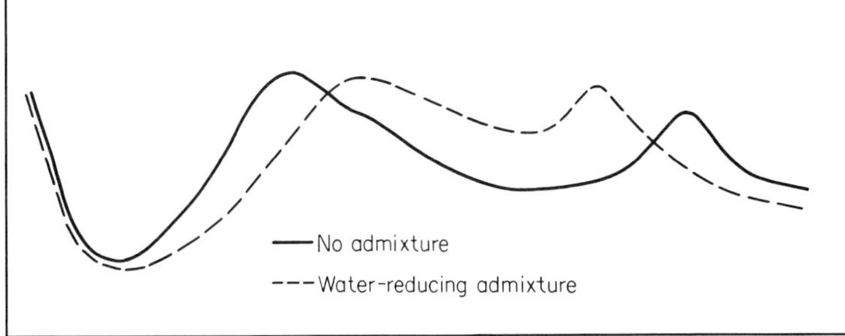

Fig. 1.27 The effect of a water-reducing admixture on the heat evolution of cement.

water at various times in the presence or absence of gypsum and calcium lignosulphonate (CLS). Table 1.11 illustrates the results obtained [69]. These results reinforce the suggestion that calcium lignosulphonate merely delays the conversion of C_3A to its stable hydrated form C_3AH_6 in a similar manner to gypsum. In addition, the rate of formation and conversion of ettringite to monosulphate is also delayed. This work also suggested that the hexagonal plates of C_4AH_x hydrates were thinner in the presence of calcium lignosulphonate, which may be responsible for an increase in strength observed.

(c) As far as the final hydration products of ordinary Portland cement are concerned, there is an indication from isothermal calorimetry [70] that there is very little difference in the presence or absence of a calcium lignosulphonate water-reducing admixture. In this work, the heat evolved per unit of water incorporated into the hydrate has been determined for two cements with the results shown in Fig. 1.28. It can be seen that the relationship between the amount of heat evolved and the amount of water combined with the cement is maintained whether the admixture is present or not. This work also indicated that the retardation in the early stages is compensated for at later times by an acceleration.

(d) In the presence of calcium lignosulphonate [71], the calcium silicate hydrate gel from the C_3S and C_2S phases tends to have a greater proportion of the crumpled foil morphology type than the corresponding system without the admixture. This observation tends to be made only at high concentrations of calcium lignosulphonate in the region of two to four times the normal dosage. This morphological type is believed to have greater drying potential than the other hydrate types of C–S–H gel and may be the reason why occasionally high drying shrinkages are observed in the presence of water-reducing admixtures.

(e) The addition of calcium lignosulphonate or a hydroxycarboxylic acid

Table 1.11 Effect of calcium lignosulphonate on C_3A reaction (after Chatterji)

	C_3A+H_2O			C_3A+Gypsum			C_3A+Gypsum+0.2% CLS			C_3A+0.2% CLS		
	1 day	14 day	3 month	1 day	14 day	3 month	1 day	14 day	3 month	1 day	14 day	3 month
C_3A	80%	20%	0	60%	30%	20%	70%	60%	30%	80%	60%	10%
Gypsum	—	—	—	20%	0	0	20%	10%	0	—	—	—
C_2AH_8	10%	trace	trace	—	trace	10%	0	0	10%	10%	10%	trace
C_4AH_x	10%	20%	20%		10%	10%	0	0	10%	10%	20%	30%
C_3AH_6	trace	60%	80%	—	—	—	—	—	—	—	10%	60%
Ettringite	—	—	—	20%	0	0	10%	20%	0	—	—	—
Monosulphate	—	—	—	trace	60%	60%	trace	10%	50%	—	—	—

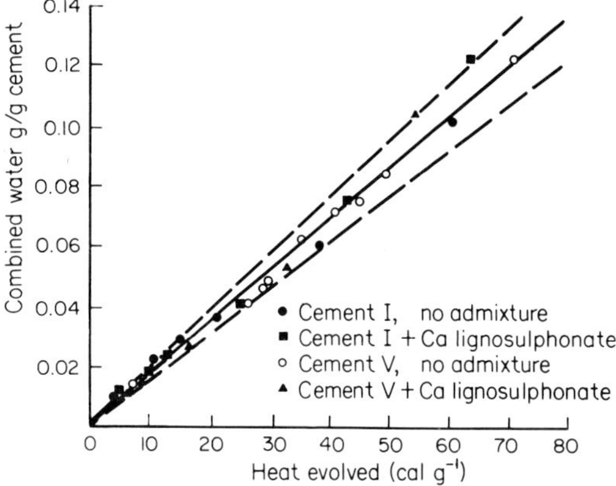

Fig. 1.28 The heat evolved from hydrating cement with and without the addition of calcium lignosulphonate as a function of combined water (Khalil).

water-reducing admixture [63] to tricalcium silicate systems can alter the morphology of the resultant calcium hydroxide to produce irregular crystals differing dramatically from the normal hexagonal form. However, this is not always observed and in one study only one calcium lignosulphonate out of four altered the morphology of the calcium hydroxide crystals. In addition, there seems to be an effect on the number of crystals per unit volume although again this can be either increased or decreased depending on the type of material used.

1.3.4 Interpretation in terms of a mode of action

It is considered that water-reducing admixtures operate by chemically reacting through their acidic and hydroxyl functional groups with predominantly the early hydration products of the C_3A phase and to some extent with the C_3S initial-hydration products to form a monomolecular layer of admixture at the water–cement interface. This monomolecular layer will be associated with a sheath of water molecules to form a total barrier some tens of angstroms thick which will prevent close approach of cement hydration particles and reduce the effect of the van der Waals forces of attraction normally operating at close proximity. Thus the interparticle friction in the system is reduced so that the energy required to induce flow into the system is also reduced. An alternative and possibly in some cases concurrent mechanism is that the adsorbed molecules retain their ionic nature resulting in initial repulsion of the particles by electrostatic means.

The presence of the admixture at the surface, depending on the forces between the admixture and the surface will impose an additional barrier to the diffusion of hydration products, therefore increasing the length of the dormancy period. Because of the introduction of competitive material for sites, particularly on the C_3A phase, the reactions between C_3A and gypsum are slightly modified in the early stages although subsequently the overall products of reaction are very similar.

1.4 The effects of water-reducing admixtures on the properties of concrete

The properties of concrete can be considered in terms of a number of stages:

The *initial plastic* state of the fresh concrete subsequent to the mixing process, where properties such as the air content, density and workability are normally measured by relevant standard tests, and utilized as a means of control of production. The magnitude of these properties is affected by the addition of water-reducing admixtures, either intentionally or as side effect, which could result not only in a change in the characteristics in the plastic state, but could also be reflected in changed properties in the hardened state.

The *later plastic* state when the concrete may be transported, handled and placed and where changes in properties such as workability and the ability of the mix to resist segregation and bleeding may affect these operations.

The *hardened* state at a relatively early date, usually 28 days, when the mechanical properties such as compressive and flexural strength and stiffness are used as a basis of structural design.

The *subsequent hardened* state during the life of the structure where the concrete material must fulfil its structural or aesthetic role without deterioration. It is important that the durability of the concrete should not be adversely affected by the presence of a water-reducing admixture.

1.5 The effects of water-reducing admixtures on the properties of plastic concrete

1.5.1 Air entrainment

During the mixing of concrete, the 'folding' action of the mixing sequence causes air voids to be formed in the system, which in normal concrete would be reduced by the mechanical forces used in placing the concrete, leaving perhaps up to 1.5% air by volume trapped under aggregate particles. In UK practice it is generally considered undesirable to allow air contents to rise much above this level for structural concete, because of the effect on compressive strength. In North America, where air-entrained concrete is more widely used, the use of those water-reducing admixtures which have a tendency to increase air contents will necessitate the reduction of the dosage of the air-entraining agent, often by as much as 50%. On the other hand, certain superplasticizers, particularly those based on the melamine and naphthalene formaldehyde sulphonates, when used to reduce water–cement ratio, require a significant increase (up to tenfold) in dosage of the air-entraining agent to achieve normal level of entrained air [72].

The presence of a water-reducing admixture can alter the air content of concrete, either as a deliberate measure (the air-entraining water-reducing admixtures) or as a side effect of the material in lowering the surface tension of the aqueous phase.

The amount of air entrainment obtained will obviously vary according to the type and quantity of admixture used, as well as mix design parameters, but in general at normal dosage levels, in a 50 mm slump sand/gravel mix of 300 kg m^{-3} cement content the changes in air content shown in Table 1.12 will be observed. Where the water-reducing admixture has been added to produce a concrete of high workability, for those materials which result in an increase in the air content, approximately 1% more air will result.

The presence of entrained air will, of course, be reflected in a reduced density in the plastic and hardened stage and its effect on subsequent properties of the hardened concrete will be discussed later.

Table 1.12 Air entrainment by water-reducing admixtures

Category of water-reducing admixture	Chemical type	Additional air content (% by volume)	Reference
Normal	Lignosulphonate	0.4–2.7	[77–79]
	Lignosulphonate + tributyl phosphate	0.3–0.6	[73]
	Hydroxycarboxylic acid	−0.2–0.3	[23,78]
Accelerating	Lignosulphonate + $CaCl_2$ or formate	0.3–0.5	[78]
	Hydroxycarboxylic acid + $CaCl_2$	0.8–1.6	[19]
Retarding	High sugar lignosulphonate	1–2	
	Hydroxycarboxylic acid	0	[77]
	Hydroxylated polymer	−0.2–0	[23]
Air-entraining	Lignosulphonate + surfactant	0.9–2.6	[74,75]
	Hydroxycarboxylic acid + surfactant	3–5	[76]
Superplasticizers	Modified lignosulphonates	1–2	
	Sodium naphthalene sulphonate formaldehyde condensates	1–1.5	[76]
	Sodium melamine sulphonate formaldehyde condensates	−0.1 to −0.25	[40]

1.5.2 Workability

The ease with which concrete can be deformed by an applied stress is known as the workability of the concrete and is measured by standard tests such as compacting factor, VeBe or slump under arbitrarily chosen conditions of sample preparation and magnitude of applied stress. The amount of deformation obtained under standard conditions would depend on the volume fraction of the aggregate and the shear resistance or viscosity of the cement paste. The effect that water-reducing admixtures have on the cement paste viscosity has been described earlier, but other factors can alter their effect on concrete such as the lubrication of aggregate particles in higher aggregate/cement ratio mixes, etc. [80].

When a normal, accelerating, or retarding water-reducing admixture is utilized to increase the workability of a concrete mix by direct addition, it would be reasonable to assume that the extent of the effect would be markedly affected by changes in mix design parameters such as cement content, aggregate size, shape and grading, and the water–cement ratio. A study of many hundreds of results, however, indicates that this is not the case and Fig. 1.29 illustrates the relationship between initial and final slump for water-reducing admixtures at normal dosage levels. The hydroxycarboxylic

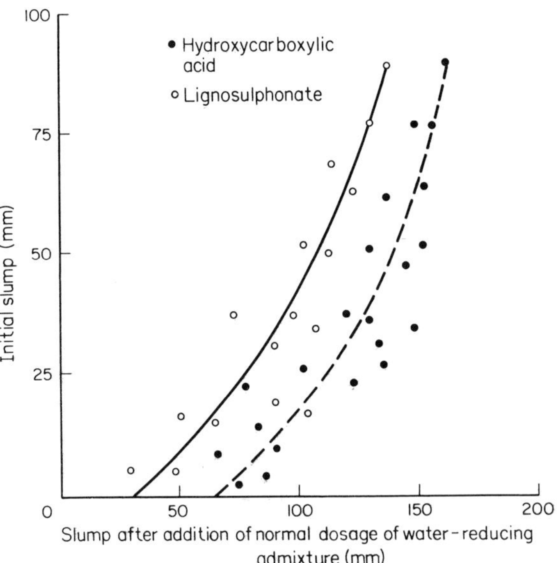

Fig. 1.29 The relationships between initial slump and the slump after the addition of water-reducing admixtures.

acid type appears to be generally superior to the lignosulphonates in increasing the value of slump, and this difference is maintained over the initial slump of 0 to 100 mm. This non-dependence of mix design parameters on the effect of water-reducing admixtures is perhaps less surprising when it is considered that factors such as wetting and adsorption of aggregates, attrition between aggregate particles, and sufficient excess water to achieve the required slump, have already been taken into consideration during the developments of the initial mix design to produce the relevant workability. Therefore the effect of water-reducing admixture is above and beyond these requirements and leads to approximately the same increase in slump across the initial slump range.

This independence of efficiency in relation to mix design parameters is only true with regard to workability increases; where a concurrent change in water–cement ratio is made, a number of variables must be considered and will be discussed later.

The increase in workability obtained is, or course, a function of the dosage of admixture used and this is illustrated in Fig. 1.30 for lignosulphonates and the hydroxycarboxylic acid material. It will be appreciated that considerable retardation would be obtained at the higher dosage levels.

It was explained earlier that superplasicizers operate in the same way as normal water-reducing admixtures, but because of the higher dosages used, the increase in workability is more dramatic. In addition, of course, the chemical materials used in their formulation do not signicicantly affect the setting or hardening rate of the concrete. Extreme workability produced by the addition of a superplasticizer requires a test other than slump or VeBe,

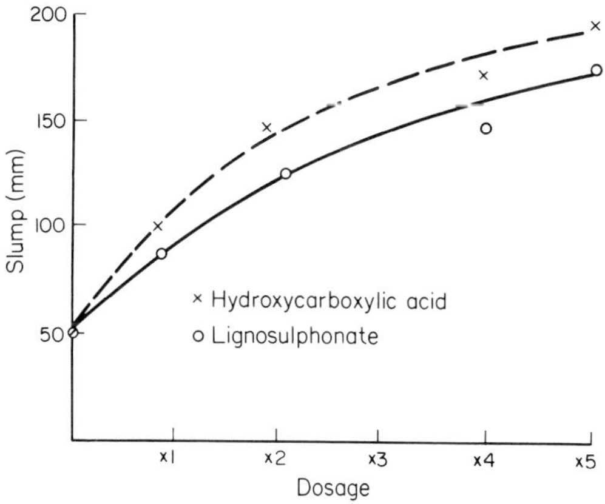

Fig. 1.30 The effect on slump of varying addition levels of water-reducing admixtures.

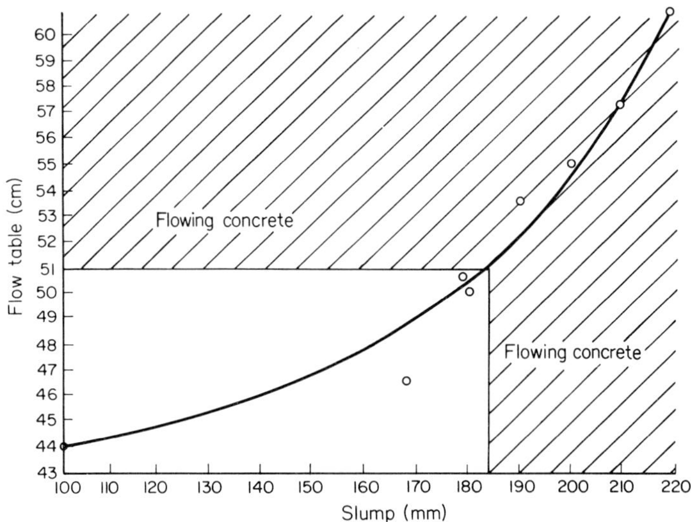

Fig. 1.31 The relationship between slump and flow table spread of concrete containing a superplasticizer.

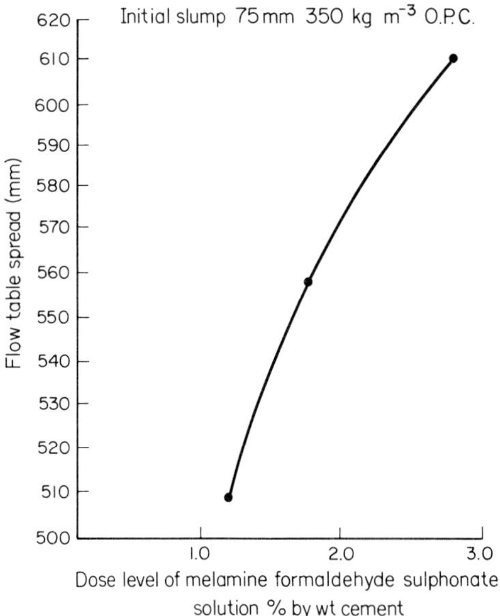

Fig. 1.32 The effect on the flow table spread of various addition levels of a superplasticizer.

and the test utilized is a slightly modified form of the German DIN 1048 standard called the 'flow table test'. This is recorded in cm spread of a cone of concrete compacted under standard conditions. A relationship between slump and the flow table spread is shown in Fig 1.31 and it can be seen that at the high slump values, the normal British standard slump would not be sensitive enough.

The workability of superplasticized concrete is dosage dependent, and typical results are shown in Fig. 1.32. The required dosage to obtain good cohesive flowing concrete of the required workability can be related to the initial slump prior to addition and, for a typical mix, results are shown in Fig. 1.33 [81]. This indicates that either the initial slump or the addition level can be used as variables to give flowing concrete conforming to DIN 1048 specification.

Workability control

The relationship between water–cement ratio and workability for mixes containing water-reducing admixtures in comparison to mixes not

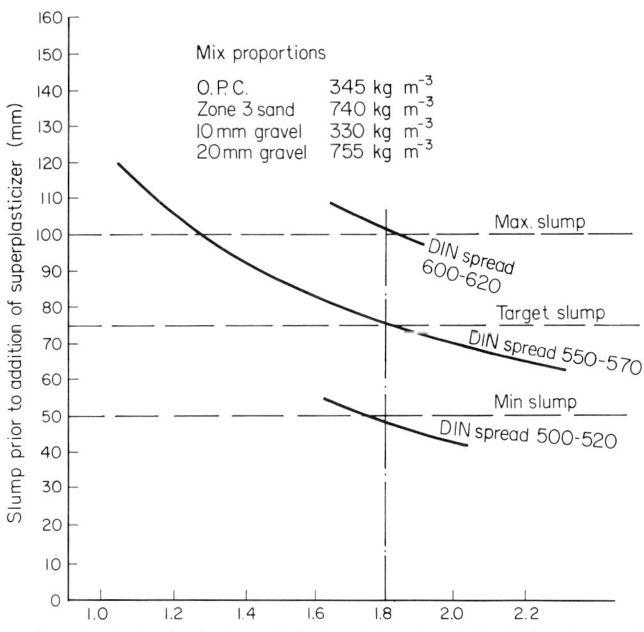

Fig. 1.33 The relationship between initial slump, the flow table spread and the addition level of a superplasticizer.

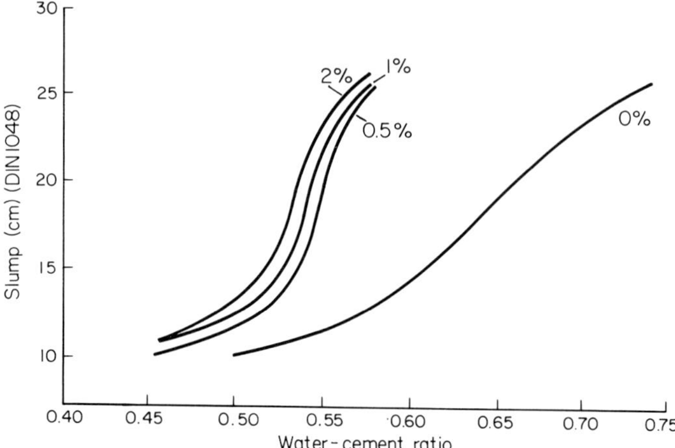

Fig. 1.34 The slump of mortars as a function of the water–cement ratio and the amount of melamine formaldehyde resin (Sasse).

containing them can be studied by consideration of Figs 1.34 and 1.35 which are for mortars containing a superplasticizer of the melamine formaldehyde type [82] and an undisclosed normal water-reducing admixture [83], respectively.

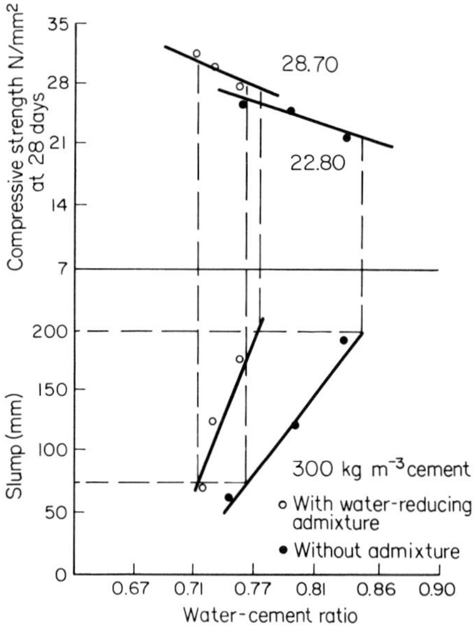

Fig. 1.35 The relationship between the slump and the water–cement ratio for mixes with and without a water-reducing admixture (Howard).

These figures illustrate that a given range of workability can be obtained over a smaller range of water–cement ratios for an admixture-containing mix. In practice this can mean that the normal variabilities in water added to the mix produce a wider range of slump values which is not conducive to accurate control of workability. However, this effect can be considered as beneficial in allowing regain of workability by addition of further water with the minimum effect on concrete quality in terms of strength. This effect has been studied more recently [84] as a means of re-tempering concrete mixes which have been subjected to prolonged mixing at elevated temperatures resulting in loss of slump. It was found that the amount of water required to regain the original workability was reduced by up to 20% when a water-reducing admixture was present in the mix in comparison to a control concrete not containing an admixture.

1.5.3 Workability loss

Concrete is judged for its suitability and quality for a given set of mix proportions by its workability, usually in terms of the slump. Once the required workability of the concrete has been attained there will be

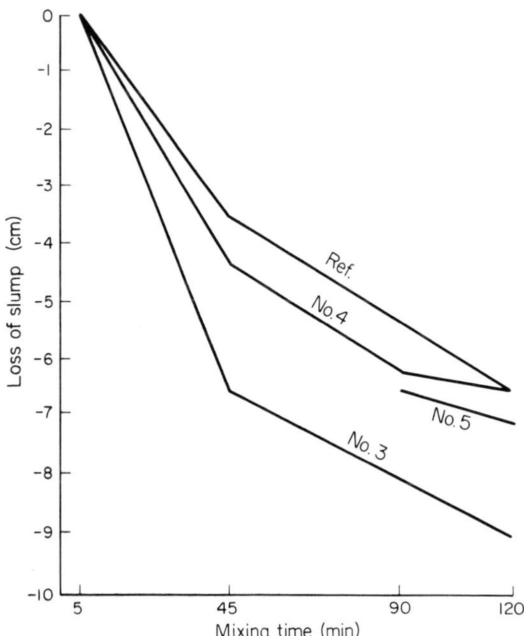

Fig. 1.36 The loss of slump with time (Ravina).

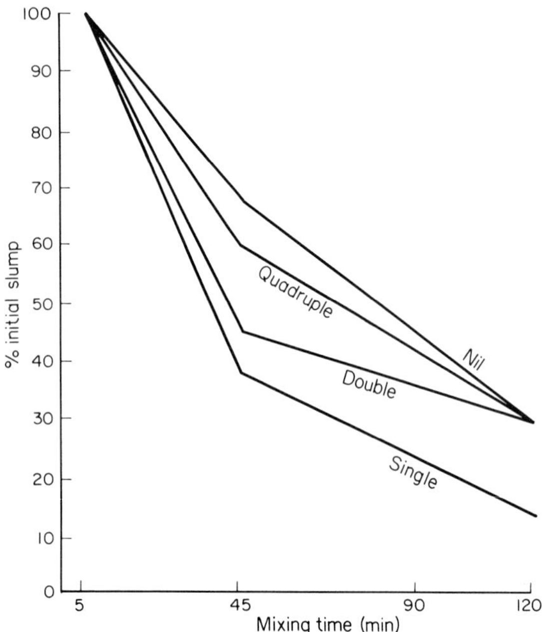

Fig. 1.37 The effect of different dosage levels on loss of slump (Ravina).

progressive loss of workability with time as the hydration process proceeds. This process continues through the mixing, discharging, handling, placing and vibrating and any changes in the rate at which workability is lost can affect any or all of these steps. The loss of workability generally appears to be more pronounced with mixes containing water-reducing admixtures and is illustrated in Fig. 1.36 [84]. All mixes were designed to initial slump (ASTM) of 10 ± 1 cm and had a cement content of 300 kg m^{-3}. An increase in the dosage apparently reduces the slump loss as shown in Fig. 1.37 [84].

Both Figs 1.36 and 1.37 illustrate the loss of slump from those mixes designed to initial slump equivalent to a mix containing no admixture. However, when the water-reducing admixture has been used to increase the workability by a straight addition, although the rate of slump loss is still greater in the case of the admixture-containing mixes, the high workability is maintained for a longer time as shown in Fig. 1.38 [84]

Similar results are obtained for hydroxycarboxylic acid based retarding water-reducing admixture and are shown in terms of loss of workability measured by BS1881 slump test and by the VeBe in Fig 1.39. The general conclusion can be reached that the use of retarding water-reducing admixtures to increase the initial workability so that the initial rate of the slump loss is compensated for, will prolong the time available for the

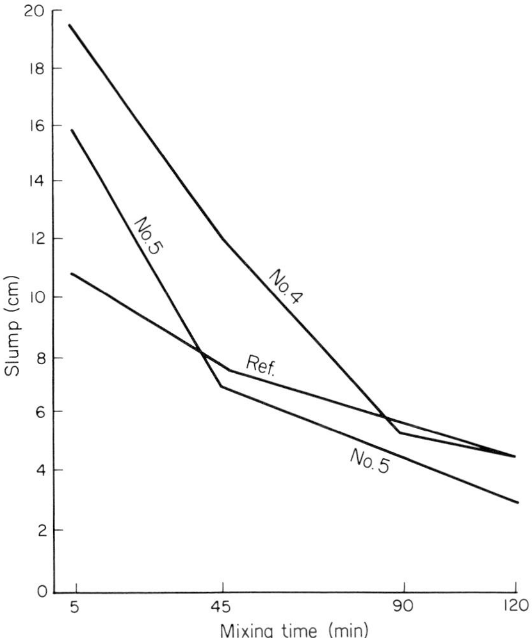

Fig. 1.38 The loss of slump with time when straight addition of a water-reducing agent is made (Ravina).

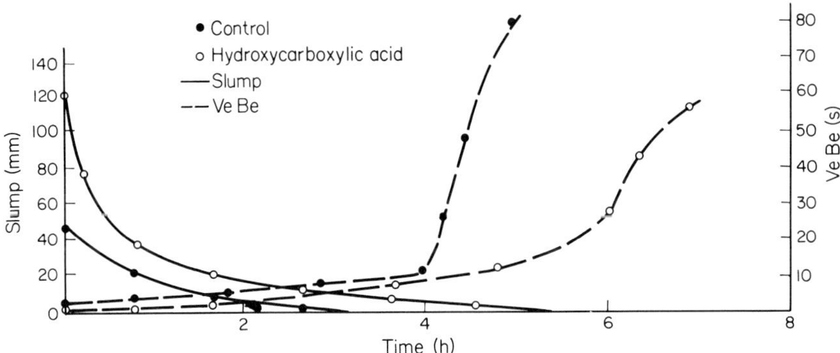

Fig. 1.39 Changes in slump and VeBe values for concrete containing straight addition of a hydroxycarboxylic acid based water-reducing agent.

transporting, handling and placing of concrete. Even when these types of materials are used to produce concrete of normal workability, it is generally found that the increased slump loss would cause no problems in normal concrete production unless particular circumstances such as hot weather or long hauls are involved. In these cases the amount of water required to

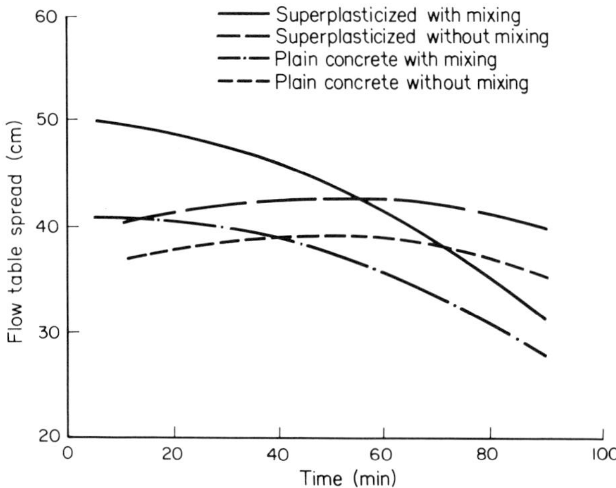

Fig. 1.40 Flow table–time relationships for plain and superplasticized concretes.

correct the loss of slump is reduced in the presence of a water-reducing admixture. This statement applies to the majority of cases, but there have been instances of severe loss of slump, which have hampered concreting operations and it has been suggested [85] that this is more likely to occur in high alkali cements. The problem is minimized by the addition of the admixture after the mixing ingredients have been given an initial mixing cycle of 2 min.

A similar effect of loss of workability is noted in the case of superplasticized concrete of the sulphonated melamine formaldehyde or sulphonated naphthalene formaldehyde types. Often the phenomenon is more pronounced because of the extreme initial workability obtained. Fig. 1.40 is a typical flow table against time relationship of superplasticized flowing concretes [86] and also gives some indication of the effect of agitation.

The initial workability and subsequent workability loss of superplasticized concrete is also a function of the age of the concrete when the addition of the admixture is made. This is illustrated in Fig. 1.41 [87].

If the superplasticizer is used to produce high strength concrete of normal workability utilizing low water–cement ratios, the slump loss is again considerably increased, as shown in Fig. 1.42 [82]. Therefore in designing concrete of this type some allowance should be made for subsequent slump loss.

The workability loss of superplasticized concrete at various temperatures has been studied and additions of retarders made to both naphthalene and melamine formaldehyde sulphonate based superplasticizers in order to

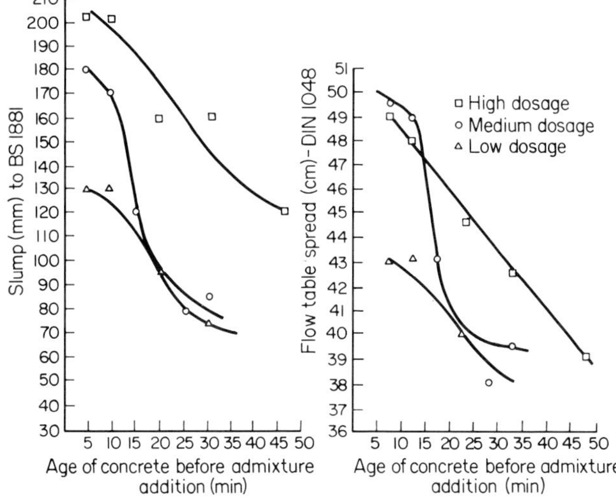

Fig. 1.41 The effect of age of the concrete on the capability to be rendered flowing by the addition of a superplasticizer.

extend the period of workability. Table 1.13 shows the effect on workability loss at various temperatures of concrete containing a specially modified naphthalene based superplasticizer [88] whilst Fig. 1.43 gives graphical results for the slump loss of cencrete containing modified and unmodified melamine based superplasticizers [89].

In the practical application of flowing concrete, because of the rapid loss

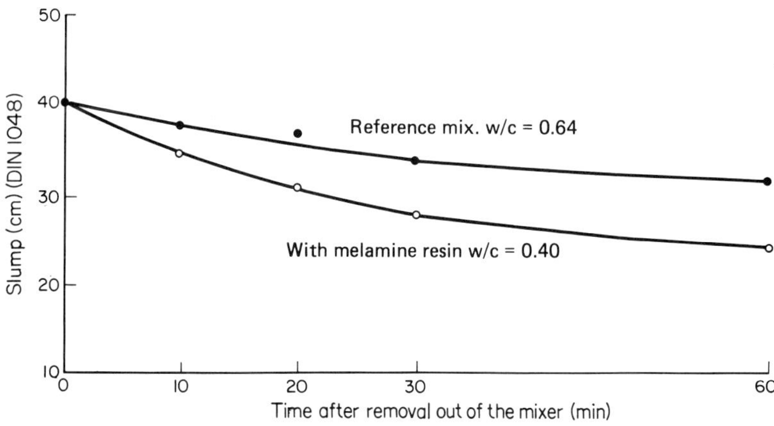

Fig. 1.42 The workability of superplasticized high strength concrete as a function of time (Sasse).

Table 1.13 The slump loss of superplasticized concrete at various temperatures (Collopardi)

Time (h)	Slump of superplasticized concrete (mm)		
	4 °C	21 °C	42 °C
0	220	220	210
0.5	205	200	195
1	210	195	185
2	210	200	150
4	185	140	30

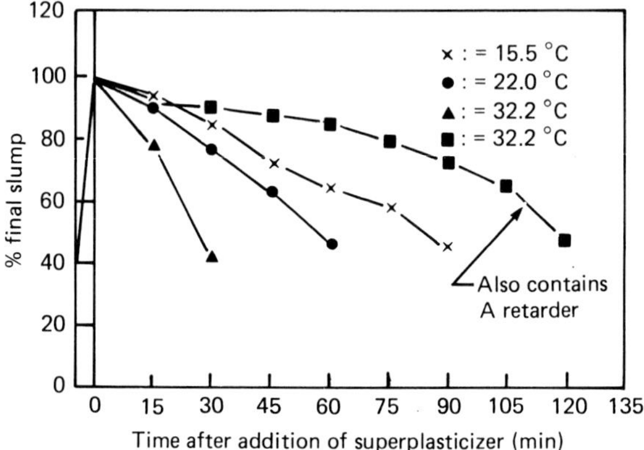

Fig. 1.43 The slump loss of a melamine formaldehyde sulphonate superplasticized concrete at various temperatures (Mailvaganam).

of the extreme workability, the admixture should be added on site just prior to placing (see Section 5.4). Alternatively, it is reported [90] that instead of making the addition in one dose, an incremental addition can be made which prolongs the workability; this is illustrated in Fig. 1.44.

1.5.4 Water reduction

The most widely used application of water-reducing admixtures is to allow reductions in the water–cement ratio whilst maintaining the initial workability in comparison to a similar concrete containing no admixture. This, in turn, allows the attainment of a required strength at lower cement content to effect economies in mix design.

The amounts of water reduction possible depend on numerous factors and these are summarized below.

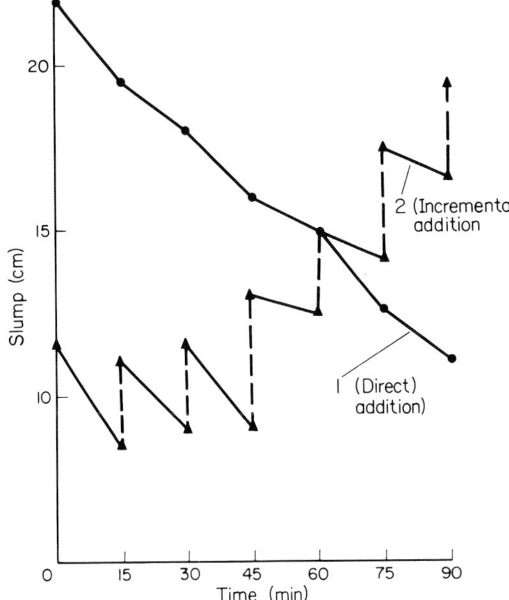

Fig. 1.44 Workability can be maintained for a longer time by incremental addition of a superplasticizer.

(a) The aggregate–cement ratio

The efficiency of water-reducing admixtures, and their relative usefulness are dependent on the aggregate–cement ratio. Hydroxylated polymer and hydroxycarboxylic acid types are more effective than lignosulphonate based materials at higher cement contents (lower aggregate–cement ratios), whilst the lignosulphonate materials are generally preferred for the lower cement contents (high aggregate–cement ratio) mix designs. Typical comparative data are shown in Fig 1.45. It can be seen that the water-reducing admixtures are most effective at an aggregate–cement ratio in the region of 6.5 to 7.0 in these mixes.

(b) Designed workability

The higher the required workability, the greater is the reduction in water–cement ratio when an addition of a water-reducing admixture is made. Thus for a typical 300 kg m^{-3} concrete with natural gravel aggregates and with a zone 3 sand, the typical values in Table 1.14 would apply for a normal addition level of a lignosulphonate water-reducing agent.

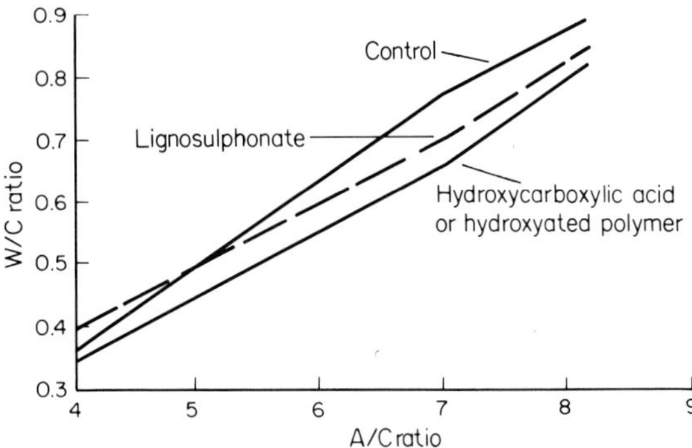

Fig. 1.45 Reductions in water–cement ratio as a function of aggregate–cement ratio for lignosulphonate and hydroxycarboxylic acid based water-reducing agents.

Table 1.14 Water reduction by water-reducing agents as a function of workability

Designed slump (BS 1881) (mm)	% reduction in w/c ratio
50	5–8
75	8–10
100	10–12
150	12–15

(c) Addition level

It is possible to vary the addition level of water-reducing admixtures when an increase in dosage level will generally produce an increase in the amount of water which it is possible to remove from the mix proportions whilst maintaining the required slump. Typical values are shown in Table 1.15 for an aggregate–cement ratio of 5.85:1 and a slump of 50 mm.

When superplasticizers are used to effect reductions in water–cement ratio, much larger decreases in the water required are obtained. The effect is dependent on the amount added as shown in Fig. 1.45 [87].

Similar results for melamine formaldehyde based materials have also been reported [91] and these data are given in graphical form in Fig. 1.46. It can be seen that by the use of superplasticizers considerable reductions in water–cement ratio can be obtained to produce much higher strength concrete as shown in the next section.

The amount of water reduction possible is also a function of the way in which an admixture is added to the concrete; if a period between mixing with water is allowed prior to the addition of the admixture, greater adsorption of the admixture on to the initial hydrates is obtained and a higher workability

Table 1.15 Effect of addition level of water-reducing admixtures on the water reduction

Water-reducing admixture type	Addition level	W/c ratio
None		0.55
Lignosulphonate	Normal	0.51
	2 × normal	0.49
	5 × normal	0.47
None		0.55
Hydroxycarboxylic acid	2 × normal	0.48
	5 × normal	0.46

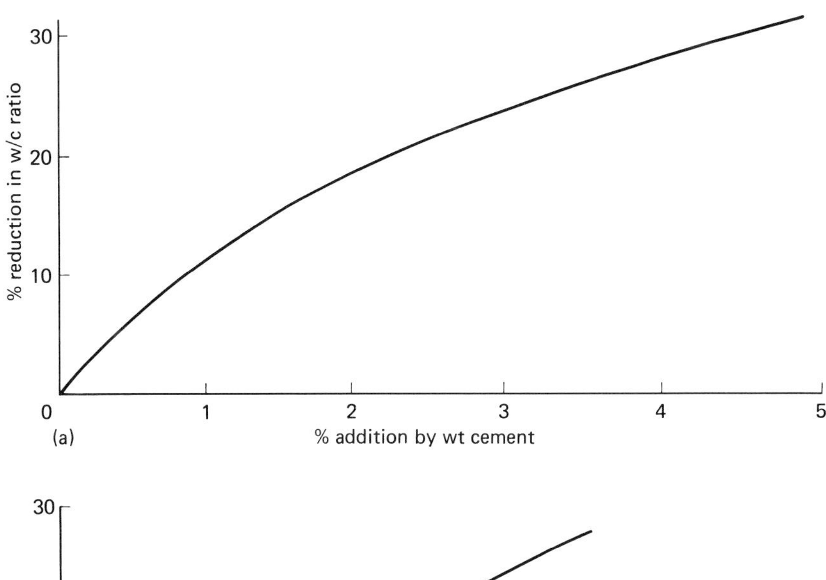

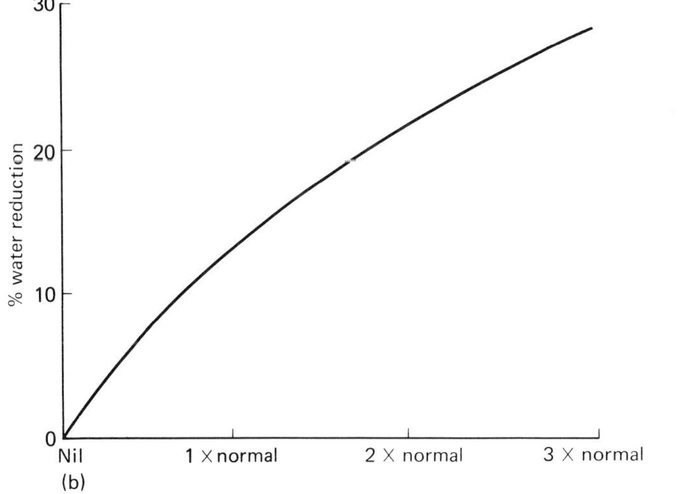

Fig. 1.46 Typical reductions in water–cement ratio obtained by two types of superplasticizers [(a) after Hewlett (b) after Van der Zanden].

or alternatively a greater reduction in water–cement ratio is obtained as can be seen from Table 1.16 [92]. This information is, on first sight, in contradiction to the data for superplasticizers shown in Fig. 1.41 where the ageing of the concrete prior to superplasticizer addition appears to reduce the ability to produce flowing concrete. However, the data shown in Fig. 1.41 start at 5 min. and it has been found with superplasticizers that if they are added to the mix water without some prior hydration period, the effect is very much reduced.

Table 1.16 Effect of varying the point of addition on workability and/or water reduction (after Dodson)

Method of addition of retarder (0.225% calcium lignosulphonate by wt cement)	W/c ratio	Slump (mm)	% Water reduction
No retarder added	0.59	100	—
Added with mix water	0.55	88	6.8
Addition delayed 2 min	0.55	163	6.8
Addition delayed 2 min	0.51	81	13.6

(d) Cement characteristics

In the case of lignosulphonate water-reducing agents, the effectiveness in reducing the water–cement ratio diminishes with an increase in either the the C_3A or alkali content. This is not an area that has been well quantified in the literature, but experience has shown that the effect can be considerable and in a comparative experiment with three cements varying in C_3A content from 9.44 to 14.7% in comparable mixes, the percentage water reduction for a calcium lignosulphonate based material varied from 4 to 10% to achieve a similar level of workability. There is some evidence that the hydroxycarboxylic acids are less dependent on cement variables than the calcium lignosulphonate based materials.

1.5.5 Setting characteristics of fresh concrete containing water-reducing admixtures

The effect of mix and environmental factors on the setting characteristics of concrete in the presence of admixtures which exert a retarding influence are discussed later. As a generalization, Table 1.17 can be used as a guide for a 300 kg m^{-3} ordinary Portland cement concrete mix having a slump in the range 50 to 100 mm and the initial setting time (measured by Proctor needle, ASTM C403) in the region of 7 to 8 h.

In view of the differences in initial setting time, most calcium

Table 1.17 Extension of initial setting time by various water-reducing admixtures

Admixture type	Dosage	Extension of initial setting time (h) at 20 °C
High grade calcium lignosulphonate (normal water-reducing)	1 × normal	4
	2 × normal	10
	3 × normal	16
Hydroxycarboxylic acid type (retarding water-reducing)	1 × normal	6
	2 × normal	12
	3 × normal	17
Lignosulphonate/calcium chloride (accelerating water-reducing)	1 × normal	−1
Sodium lignosulphonate (normal water-reducing)	1 × normal	0.5
	2 × normal	2.0
	3 × normal	3.5

lignosulphonates and hydroxycarboxylic acid based materials would extend the time available for transport, handling, placing and finishing of concrete, whereas the materials based on sodium lignosulphonate will have only a marginal effect. Some ligosulphonates contain a small amount of triethanolamine to overcome the retardation effect of the major component but, as described later, this can be undesirable where possible volume deformation changes are important in the resultant concrete.

1.5.6 The stability of fresh concrete containing water-reducing admixtures

The stability of the concrete mix can be considered in terms of its 'cohesion', which is a subjective term used to describe its ability to maintain a homogeneous appearance when subjected to applied stress. Lack of cohesion leads to segregation of the mix components into layers relevant to their densities. A further term associated with mix stability is that of 'bleeding', which is the movement of water to the surface of the fresh concrete. This phenomenon can occur either in isolation or as a manifestation of segregation. Bleeding in excess is normally considered to be undesirable because of the dangers of water runs at the shutter/concrete interface and cracking due to plastic settlements, and there is also the possibility of adverse effect on the concrete–reinforcement bond due to the collection of water beneath the steel.

(a) Cohesion

There is little published data on the cohesion of mixes containing water-reducing agents presumably because of the absence of a truly quantitative

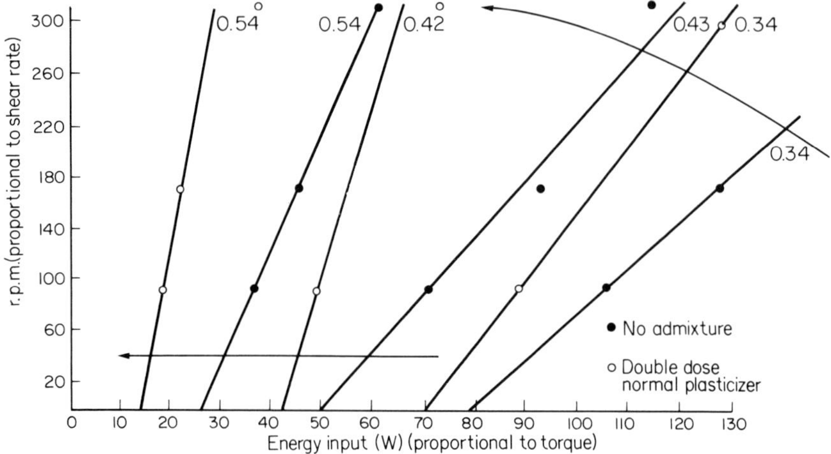

Fig. 1.47 Rheology of concrete containing a normal plasticizer (300 kg m^{-3}) (Hewlett).

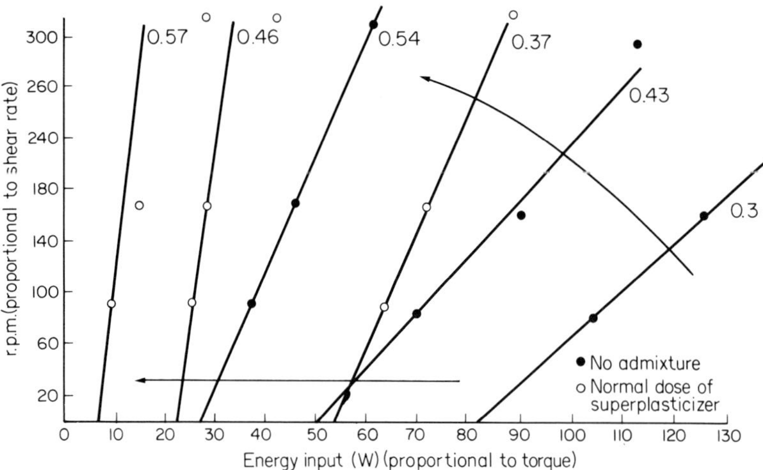

Fig. 1.48 Rheology of concrete containing a superplasticizer (Hewlett).

method of measurement. A general observation is that when a water-reducing admixture is used to produce a higher strength concrete at a reduced water–cement ratio, the concrete appears to be more cohesive. In addition, it is often noted that increases in workability produced by the addition of a water-reducing admixture can be made without the loss of cohesion associated with redesigning the mix at higher water content. This latter point is particularly important in the case of the superplasticizers where it is claimed that extremely high workability can be obtained without excessive loss of cohesion.

The only quantitative data useful in this context in the published information [93] are the rheological characteristics of concrete containing a lignosulphonate water-reducing admixture, a sodium naphthalene sulphonate superplasticizer, and a control mix. Figs 1.47 and 1.48 show the rheological results and the slope of the line is a function of the workability of the concrete whilst the intercept on the energy input axis should be an indication of the cohesion or the inherent structure of the concrete. Fig. 1.49 relates the workability to the cohesion in these terms.

These limited results give some indication that it is possible to achieve high workability material without a consequential loss in cohesion by the use of water-reducing admixtures of the lignosulphonate or superplasticizer type.

(b) Bleeding

The movement of clear water to the surface of the concrete can lead to aesthetic problems of surface finish and plastic settlement and interfere with the reinforcement–concrete bond. On the other hand, promotion of bleeding can, in certain circumstances, be beneficial; in hot and windy weather conditions plastic cracking can occur when the rate of evaporation exceeds the bleeding rate. Also concrete which has shown severe bleeding is

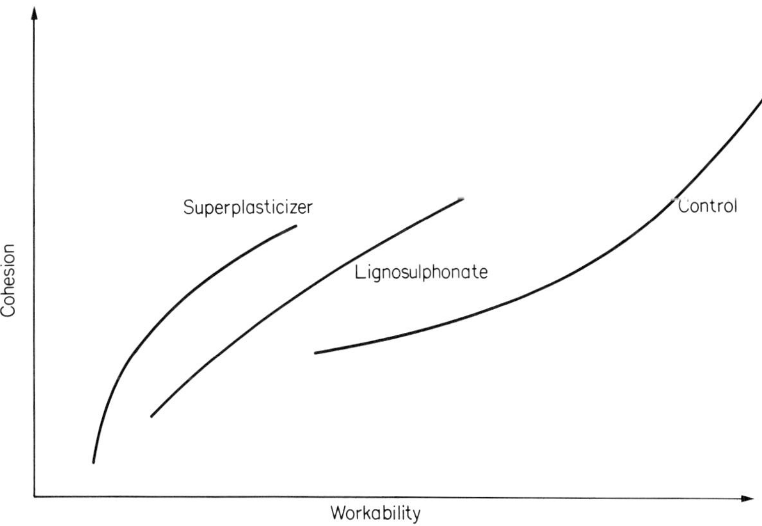

Fig. 1.49 Concretes containing water-reducing admixtures tend to have a more cohesive nature than a plain concrete at a given workability (after Hewlett).

often stronger [94] because of the reduction in water–cement ratio. In general it can be said that the addition of any water-reducing admixture or superplasticizer which does not significantly increase the air content will lead to an increase in the rate and capacity of bleeding of the plastic concrete. Lignosulphonate based materials usually lead to an increase in air content which can result in an overall decrease in bleeding, whilst normal or retarding water-reducing admixtures containing hydroxycarboxylic acid materials invariably give an increase in bleeding rate.

However, when all categories are used to reduce the water–cement ratio to the same workability as a concrete containing no admixture, this, in turn, reduces bleeding so that the net effect is an increase or decrease. It is reasonable to assume that an air-entraining water-reducing admixture based on any raw material type will not increase the bleeding rate whether used as a means of increasing workability or decreasing the water–cement ratio. Limited data on water-reducing admixtures containing calcium chloride do not indicate any increase in bleeding rates for the accelerating water-reducing admixtures.

Table 1.18 summarizes published data for a variety of water-reducing admixtures in a range of mix designs.

Superplasticizers, by virtue of their considerable deflocculating effect appear to increase bleeding significantly [96] and mix design procedures (see next section) are necessary to overcome this.

1.5.7 Mix design considerations

It is clearly not the purpose of this book to give a guide to the principles of good mix design, which is already catered for by a number of excellent books and reviews (e.g. [97–99]) but rather to set down a few points which are relevant to the use of water-reducing admixtures.

(i) When water-reducing admixtures are used at normal dosage levels to obtain a higher workability for a given concrete mix, there is no necessity to make any alteration to the mix design from that produced for the concrete of the initial lower slump. There is generally no loss of cohesion or excess bleeding even when the hydroxycarboxylic acid materials are used.

(ii) If this class of product is used to decrease the water–cement ratio, again no change in mix design will be required, although small alterations in plastic and hardened density will be apparent and should be used in any yield calculations. One interesting area is in the production of concrete of high elastic modulus. Most high modulus aggregates are obtained as crushed rocks [100] but this potential advantage is lost because of the high water demand to obtain the required workability. The use of water-

Table 1.18 Effect of water-reducing admixtures on bleeding rates and capacities

Product type	Mix details		Air content	Bleeding		Reference
				mls/m cm^{-2} surface	% total mixing water	
Lignosulphonate	Crushed limestone 310 kg m^{-3} cement	Control	1.6	0.48	14.2	[94]
	Approx 100 mm slump	Admixture	4.8	0.19	6.3	
Lignosulphonate	300 kg m^{-3} cement	Control	0.4	0.50	14.1	Own data
	Natural gravel	Na ligno	0.7	0.92	15.2	
	50 mm control	Ca ligno 1	1.2	0.22	7.8	
	No reduction in	Ca ligno 2	1.7	0.22	5.7	
	w/c ratio with admixture	Ca ligno 3	1.4	0.19	4.4	
Hydroxycarboxylic acid	310 kg m^{-3} Rounded aggregate –increased	Control	1.5	—	4.3	[94]
	workability	Admixture	1.3	—	5.5	
	310 kg m^{-3} cement Angular aggregate –reduction in w/c	Control	1.0	—	7.1	[94]
	ratio with admixture	Admixture	1.1	—	8.6	
Lignosulphonate/$CaCl_2$	300 kg m^{-3} cement	Control	1.4	0.13	3.4	[95]
	Slump 7.5 ± 0.5 cm	Admixture	4.4	0.08	1.5	

reducing admixtures can overcome this problem by reducing the paste water–cement ratio.

(iii) Economies in mix design are effected by reducing the cement content whilst maintaining the same water–cement ratio. In view of the reduction of paste volume, conflicting recommendation are made of how this should be compensated for. The following is a guide [98]:

(a) If the mix appears to be 'sticky', the paste volume being removed should be balanced by an increase in the sand content.
(b) If strength is the major consideration, the paste volume should be balanced by an increase in the coarse aggregate content.
(c) Most usually the paste volume is replaced by an increase in the total aggregate without a change in the coarse to fine proportioning. This technique has been used in many hundreds of applications, and only occasionally has it been found necessary to revert to (a) or (b).

(iv) It was shown earlier that aggregate types do not materially affect the performance of water-reducing admixtures. This is not true for cement and mixes containing special cements require particular care. Examples here are increased retardation with low C_3A cement (for example, sulphate-resistant cement) and even an almost complete reduction in expansive properties with expansive cements in the presence of water-reducing admixtures. However, pozzolans such as fly ash appear to behave normally with water-reducing admixtures.

(v) The production of flowing or self-levelling concrete using superplasticizers does require the development of a suitable mix design to minimize segregation and bleeding. It is particularly a problem with concretes of low cement content and with fine aggregates of zones 1 or 2 [91]. Table 1.19 gives changes in sand content to be used as a guide to

Table 1.19 Increases in sand content to maintain cohesion in flowing concrete

Cement content (kg m^{-3})	Sand grading (BS zones)	% increase in sand as proportion of total aggregate
300	1	Not suitable
	2	Not suitable
	3	5
	4	3
400	1	Not suitable
	2	5
	3	3
	4	0
500	1	0
	2	0
	3	0
	4	0

produce 75 mm slump concrete suitable for the addition of a superplasticizer from a concrete mix that has been developed by normal mix design techniques and meets the requirements of strength and cohesion at the 75 mm workability.

1.6 The effects of water-reducing admixtures on the properties of hardened concrete

The major physical attributes of concrete as a construction material are a high compressive strength and stiffness, an ability to protect and restrain steel and, most important of all, to retain these properties over a considerable period of time. The effects that water-reducing admixtures have on these properties can be considered from the point of view of design parameters, i.e. those properties of concrete at a relatively early age (usually 28 days) which are used for load calculations, and longer term aspects or durability.

1.6.1 Structural design parameters

The three most important properties of concrete used in calculations for load-bearing applications are the compressive strength, the tensile strength and the modulus. However, for certain applications, e.g. water-retaining structures, the permeability or porosity of the concrete will be a relevant design criterion and this is also considered here.

(a) Compressive strength

The compressive strength at 28 days of concrete containing water-reducing admixtures of the lignosulphonate, hydroxycarboxylic acid, melamine formaldehyde sulphonate and naphthalene formaldehyde sulphonate types is a function of the water–cement ratio and conforms to Abram's rule in the manner of concrete or cement paste [101] which does not contain an admixture. It is often claimed that materials of these types produce higher 28-day compressive strength for a given water–cement ratio, but the author has not found this in his own work. Typical data for British cements and aggregates are shown in Figs 1.50 and 1.51 and span a range of aggregate and mix design types for lignosulphonates, hydroxycarboxylic acid water-reducing agents, melamine formaldehyde and naphthalene formaldehyde sulphonate superplasticizers. Therefore, for materials of these types, no special consideration has to be taken into account for design purposes as far as 28-day compressive strength is concerned.

Air-entraining water-reducing admixtures require special consideration; the presence of entrained air leads to a reduction in compressive strength,

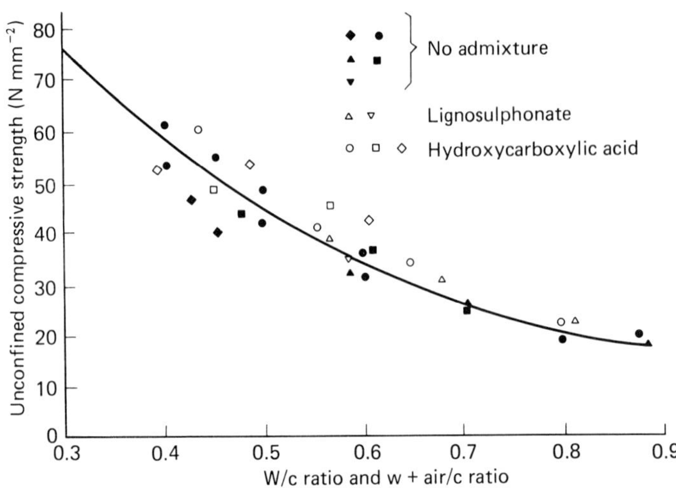

Fig. 1.50 The relationship between water–cement ratio (water and air–cement ratio) and compressive strength of concrete containing lignosulphonate and hydroxycarboxylic acid based water-reducing agents.

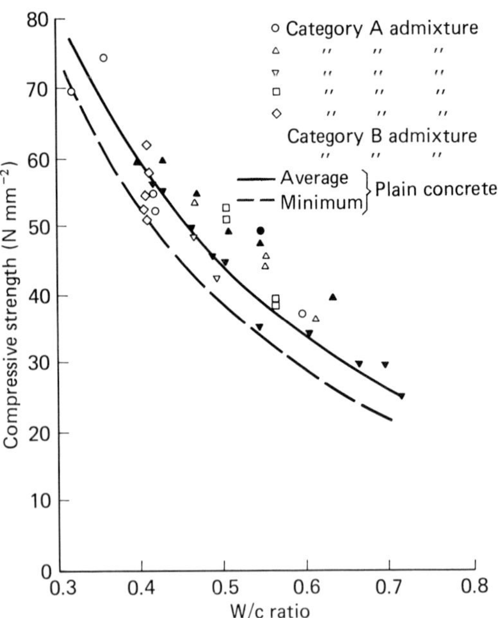

Fig. 1.51 The relationship between water–cement ratio and compressive strength of concretes containing superplasticizers.

whilst the water reduction results in a compensatory increase in strength. The effect can be quantified, however, by considering the amount of entrained air in terms of an equivalent volume of water to calculate the (air and water): cement ratio. This new factor can be used to estimate the expected strength from Fig. 1.50.

(b) Tensile strength

The tensile strength can be measured in two ways: (i) direct tensile strength from 'dumbell' specimens; (ii) splitting tensile strength from cylinders. Alternatively the flexural strength can be measured using rectangular prisms. Methods (i) and (ii) give similar values, whilst flexural strength, where the applied and resultant forces are not entirely tensile in nature, give somewhat higher values. Typical values are shown in Fig. 1.52 [102] which shows the relationship between tensile and compressive strength.

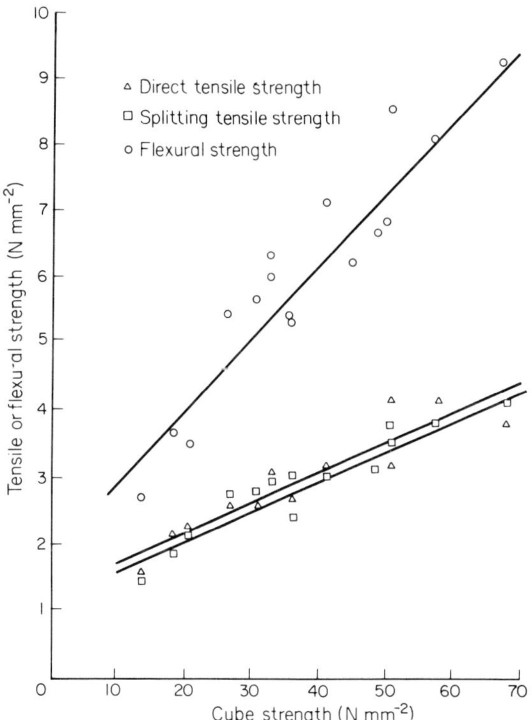

Fig. 1.52 The relationship between cube strength and tensile strength (Kromloš).

Table 1.20 Relationship between the compressive strength and the tensile and flexural strengths

Reference	Admixture type	% of compressive strength Flexural	Tensile	Average Flexural	Tensile
[103]	Hydroxycarboxylic acid	—	6.9		
		—	9.3		
		14.7	—		
[104]		14.6	—	15.2%	8.1%
		17.8	—		
[105]		15.7	—		
		13.4	—		
[103]	None	—	6.3		
		—	8.9		
[104]		15.1	—		
[105]		13.8	—	16.2%	8.8%
		16.0	—		
		16.8	10.7		
[102]		18.2	8.5		
		17.0	7.6		
[106]			10.6		
[103]	Lignosulphonate		7.1		
			7.8		
		15.2		14.9%	7.5%
		16.3			
		13.2			

Only limited data are available to illustrate the effects of water-reducing admixtures on the relationship between compressive strength and tensile strength. However, Table 1.20 summarizes the tensile, flexural and compressive strengths for some published results and also includes some comparative figures for control concretes.

It can be concluded that water-reducing admixtures of the lignosulphonate and hydroxycarboxylic acid types will not alter the relationship between the compressive strength and the tensile and flexural strengths.

(c) Modulus of elasticity

There is a paucity of recorded comparative data on the elastic modulus of concretes containing water-reducing admixtures. The one investigation of significance studied a lignosulphonate based material in corresponding mixes using five different cements [79] and the results are given in Table 1.21 as a ratio of the admixture-containing mix to the non-admixture-containing mix of similar workability and 28-day compressive strength parameters.

Table 1.21 Elastic modulus of concrete containing a lignosulphonate based water-reducing agent as a ratio of a plain mix (Tam)

Age (days)	Ratio of dynamic modulus Cement					Average
	1	2	3	4	5	
1	1.05	1.10	1.00	1.05	1.25	1.10
3	1.15	1.10	1.05	1.00	1.15	1.10
7	1.15	1.10	1.05	1.05	1.10	1.10
14	1.05	—	1.05	1.05	1.05	1.05
21	1.05	1.05	1.00	1.00	1.00	1.00
28	1.05	1.00	1.05	1.05	1.00	1.05
35	1.05	1.00	1.05	1.05	1.00	1.05
63	1.00	1.00	1.05	1.05	1.00	1.00
91	1.00	1.00	1.05	1.05	1.00	1.00
119	1.05	1.00	1.00	1.00	1.00	1.00
147	1.05	1.00	1.05	1.05	1.00	1.05
182	1.00	1.00	1.00	1.00	1.00	1.00

There are strong indications that after 28 to 35 days curing, there is little or no difference in the modulus of elasticity between the corresponding mixes, and at earlier ages the trend is towards a higher modulus.

More recent work [107] on a hydroxycarboxylic acid based material revealed the data given in Table 1.22.

Some data [108] have been compiled on concretes containing superplasticizers of the melamine formaldehyde and naphthalene

Table 1.22 Elastic modulus of concretes containing a hydroxycarboxylic acid water-reducing agent (Brookes)

Concrete mix number	Aggregate type	W/c ratio	Admixture	28-day strength (N mm^{-2})	Modulus of elasticity at 28 days (N mm^{-2})
1.1	Quartz	0.65	No	30.0	29.6
1.2		0.65	Yes	29.3	29.2
1.3		0.60	Yes	41.8	30.5
2.1		0.45	No	38.2	33.8
2.2		0.45	Yes	40.6	35.2
2.3		0.40	Yes	46.5	39.2
3.1	Limestone	0.65	No	29.2	30.5
3.2		0.65	Yes	26.7	32.9
3.3		0.58	Yes	41.2	35.9
4.1		0.43	No	47.3	40.5
4.2		0.43	Yes	46.9	37.2
4.3		0.38	Yes	52.1	42.1

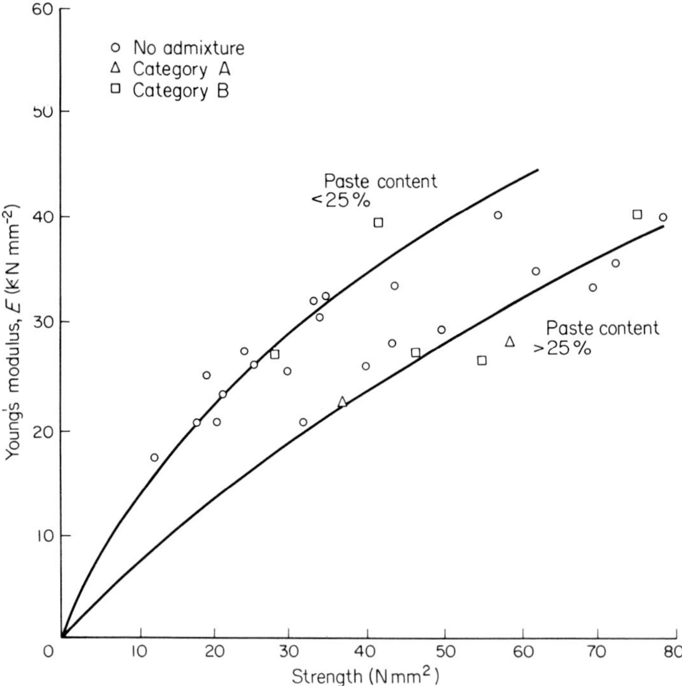

Fig. 1.53 The effect of superplasticizers on the modulus of elasticity (Blundell).

formaldehyde types and Fig. 1.53 illustrates that there is no apparent difference between concretes containing the superplasticizers and control concretes containing no admixtures.

(d) Permeability or porosity

The permeability of concrete is a guide to its durability (see Section 1.5.2) but it can also be relevant to the design of structures which are intended to withstand a hydraulic head of water or other liquid. Gross porosity is usually due to continuous passages in the concrete, due to poor compaction or cracks [91] which can be minimized by the use of water-reducing admixtures to give increased workability whilst maintaining a low water–cement ratio.

In the absence of cracks and large channels in the concrete, the permeability is a function of the paste water–cement ratio.

The graphs given in Fig. 1.54 show the logarithmic relationship between the water–cement ratio and the permeability coefficient of hardened cement

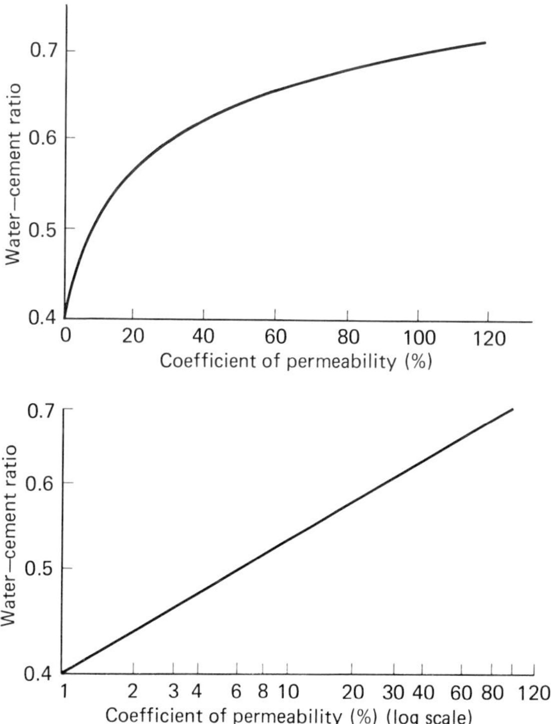

Fig. 1.54 The relationship between the permeability coefficient of concrete and its water–cement ratio.

paste. Thus concrete with a paste water–cement ratio of 0.4 will be almost impermeable. Water-reducing agents can be used to reduce the water–cement ratio, so ensuring that the permeability is kept to aminimum. Typically a concrete of paste water–cement ratio of 0.55 could be reduced to 0.50, resulting in a permeability less than half the original value.

The overall conclusion from the available data is that when a concrete mix is designed incorporating a water-reducing admixture of the normal, retarding or superplasticizer type, then the properties of the resultant concrete at 28 days will conform to the normal relationships used for concrete not containing an admixture at the same water–cement ratio.

1.6.2 Durability aspects

The durability of concrete is the ability of the material to maintain its structural intergrity, protective capacity, and aesthetic qualities over a prolonged period of time. It is important that the benefits conferred to concrete in the plastic and early hardened state by water-reducing admixtures are not negated by any adverse effect on the long term durability.

Concrete durability can be considered in terms of the following properties:

(a) The resistance to attack by aggressive liquids which would commonly be chlorides from marine environments or de-icing salts and sulphates from ground waters.
(b) The resistance to freeze–thaw cycling which may be expérienced during the winter months in many countries. This will not be a function of the average wintertime temperatures of the various countries because, in fact, the very cold environments will have only a small number of freeze–thaw cycles. In countries such as Great Britain, the winter daytime temperatures are often above 0° C and the night-time temperature below. In view of this, more freeze–thaw cycles would be experienced than in countries such as Scandinavia or North America where daytime temperatures in the winter tend to remain below 0° C.
(c) The protection of steel reinforcements. Concrete produces a layer of passivity at the steel/concrete interface and any breakdown of this can increase the chance of reinforcement corrosion. In addition, it is important that concrete be maintained in a state of low permeability to minimize the passage of moisture and air to the steel.
(d) The majority of load calculations for concrete structures are based on 28-day compressive strengths of concrete; this is based on the knowledge that concrete continues to gain in strength over the subsequent years. Any significant change in this gain of strength would

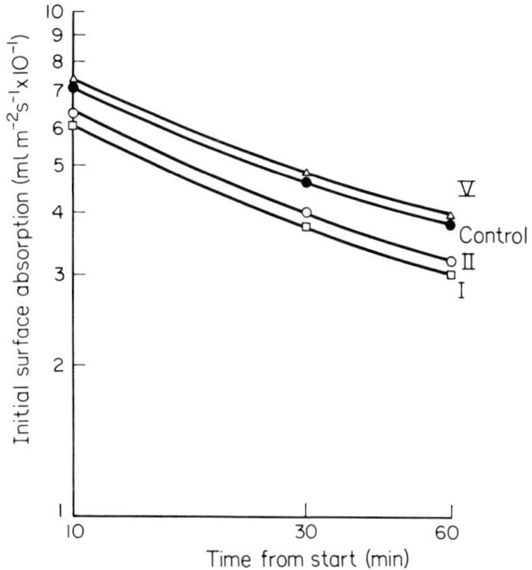

Fig. 1.55 Initial surface absorption of oven dried concretes containing lignosulphonate water-reducing agents (see Fig. 1.56 for key) (Hewlett).

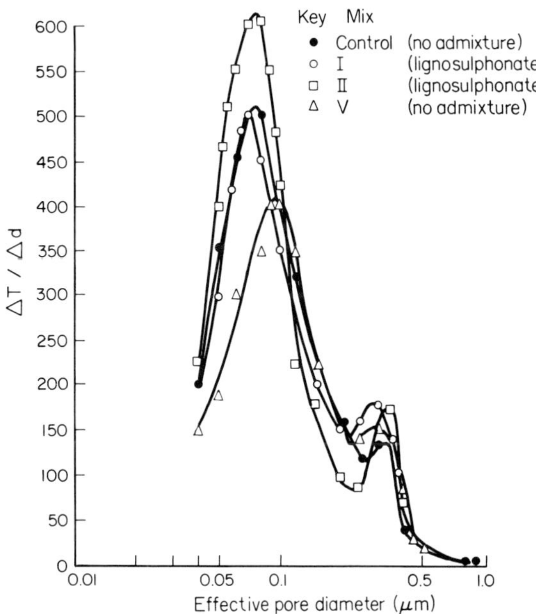

Fig. 1.56 Frequency curve of pore size distribution (Hewlett).

obviously be deleterious for the integrity of the structure and certainly any sudden change in strength characteristics could be disastrous.

(e) Concrete undergoes volume changes, particularly under drying conditions. In an unloaded state, the volume change is called shrinkage, whilst the additional volume change under an applied load is known as creep. Alterations in the rate at which a concrete shrinks and creeps due to added materials can be problematical, particularly where concretes of different volume deformation characteristics are in contact with each other, or where joints have been designed for a given rate of movement.

(a) Resistance to aggressive liquids

The deterioration of concrete under the action of materials which aggressively attack the cement matrix will be a function of the permeability or porosity of the concrete [109] and can be measured indirectly by means of the ISA test [110]. It has been shown [111] that for concrete mixes containing 255 to 300 kg m^{-3} cement designed to the same workability and 28-day compressive strength, there is no significant difference between those mixes containing no admixture and those containing a lignosulphonate plasticizer as far as the initial surface adsorption is concerned. The same concretes were also subjected to a pore size distribution assessment by means of a mercury porosimeter and these results are shown in Figs 1.55 and 1.56.

Direct measurement of the effect aggressive reagents on concrete durability appears to be confined to sea water and sulphate attack where in both areas it is recognized that the lower the water–cement ratio, the greater will be the resistance to attack and the use of a water-reducing admixture will be obviously helpful. This is confirmed by work carried out in Holland [112] and Japan [113] and a general conclusion is that a reduction in the water–cement ratio from 0.5 to 0.40, would allow a reduction in thickness of cover of the reinforcement by about 50%.

Sulphate resistance. In this area, the effect of various types of lignosulphonate and hydroxycarboxylic acid water-reducing admixture have been studied from the point of view of the effect on concrete having the same mix design but with a lower water–cement ratio in the case of the admixture-containing mixes and also a small amount of work on corresponding mixes containing lower cement contents and the same water–cement ratio and hence 28-day strength in the case of the admixture-containing mixes. Table 1.23 shows a set of results for various types of water-reducing admixture using a test method where the concrete is given periodic exposure to sulphate-containing solutions and the number of cycles to achieve a given expansion is noted together with the reduction in Young's modulus, *E*.

It can be seen that generally the lignosulphonate and hydroxycarboxylic

Table 1.23 Sulphate resistance of concrete mixes containing water-reducing agents (no mix design changes other than addition of water-reducing admixture)

Mix no.	Remarks	Accelerated sulphate test			
		0.2% expansion		0.5% expansion	
		Cycles	Reduction in E (%)	Cycles	Reduction in E (%)
	KIRWIN DAM				
80	No agent, control	226	18.4	332	57.6
81	0.25% agent L	324	20.4	449	61.9
82	0.2% agent A	354	18.7	485	47.0
	GROSS DAM				
83	No agent, control	46	6.8	65	21.6
84	0.2% agent 2	43	10.2	57	25.9
	MONTICELLO DAM				
75	No pozzolan, no agent	1150	31.8	1725*	—
76	30% pozzolan, no agent	310	60.0	554	+90.0
77	No pozzolan, agent A	780	18.5	1746*	—
78	Pozzolan, agent A	495	58.8	760	+76.0
79	Fly ash, agent A	1752	50.0	—	—
	PPT SERIES				
8	No agent, control	523	10.0	1480	—
5	0.3% agent G	550	6.7	1270	26.1
6	0.6% agent C	550	6.5	1120	20.7
7	0.25% agent D	660	8.4	1430*	—
16	No agent, control	690	6.1	1385*	—
12	0.2% agent D	662	4.3	1385*	—
13	0.3% agent D	600	7.2	1385*	—
14	0.4% agent D	895	7.1	1385*	—
15	0.5% agent D	720	14.3	1385*	—

*Cycles to time of publication without expanding to that indicated by the column heading.
Agents A: ammonium lignosulphonate solution.
D: hydroxycarboxylic acid solution.
G: calcium lignosulphonate solution.
L: calcium lignosulphonate solution.

acid type materials lead to an improvement in sulphate resistance [114]. This is borne out [115] by Russian work shown in Table 1.24 where again a direct addition of an unspecified plasticizing agent was made and the test method used here was measuring the resonance frequency of concrete specimens after different periods of immersion in a 5% sodium sulphate solution. A formula was developed to give a durability factor C*k*. This work showed that by reducing the cement content in the presence of the plasticizer, the durability in the presence of sulphate solutions was adversely affected. In view of this, it is concluded that both lignosulphonate and hydroxy-carboxylic acid plasticizers can be used to reduce the water–cement ratio of concrete mixes which would be reflected in an enhancement of the durability to sulphate attack. However, when cement reductions are made to the same

Table 1.24 Sulphate resistance of reduced water–cement ratio and corresponding mix

Specimens and type of admixture	Water–cement ratio	Durability factor *Ck*		
		10 months	20 months	28 months
Without admixture, control	0.50	1.00	0.73	0.45
With plasticizer	0.47	1.02	0.94	0.87
With plasticizer, reduced cement content	0.50	0.94	0.45	0.33

workability and strength characteristics, testing should be carried out prior to use in any sulphate-sensitive applications.

The resistance of concrete containing naphthalene formaldehyde sulphonate based plasticizers to attack by magnesium sulphate solution has been studied [88, 116]. A study of changes in weight, length and dynamic modulus of the specimens showed no adverse change in the behaviour of the superplasticized concrete in comparison with the control specimens.

In view of the known deleterious effect of admixtures containing calcium chloride and the possibility of the same effect being found with calcium formate, it is suggested that accelerating water-reducing admixtures should not be used in those areas where sulphate resistance is of importance.

(b) *Resistance to freeze–thaw cycling*

Concrete is damaged by exposure to freeze–thaw conditions due to the expansion of water in the capillaries on freezing to form ice. The expansion results in micro-cracking with a consequential loss of strength and modulus of elasticity. In addition, such concrete would become aesthetically unacceptable because of spalling at the surface, and the possibility of the ingress of water and air through the micro-cracks could lead to reinforcement corrosion. In view of this, any reduction in the water–cement ratio would be beneficial in enhancing the durability under these conditions. This is illustrated in Fig. 1.57 [117].

There is a considerable amount of recorded data on this aspect of durability, and this is summarized in Table 1.25.

(i) *Lignosulphonates.* For concrete used in dam construction, the results shown in Table 1.25 [118] have been obtained. From these results, it will be seen that a reduction in the water–cement ratios was obtained and in the large majority of cases (80% of the specimens) an improvement in freeze–thaw resistance was obtained. In fact, the average resistance of admixture-containing concrete was 39% greater than the control specimens. The ability

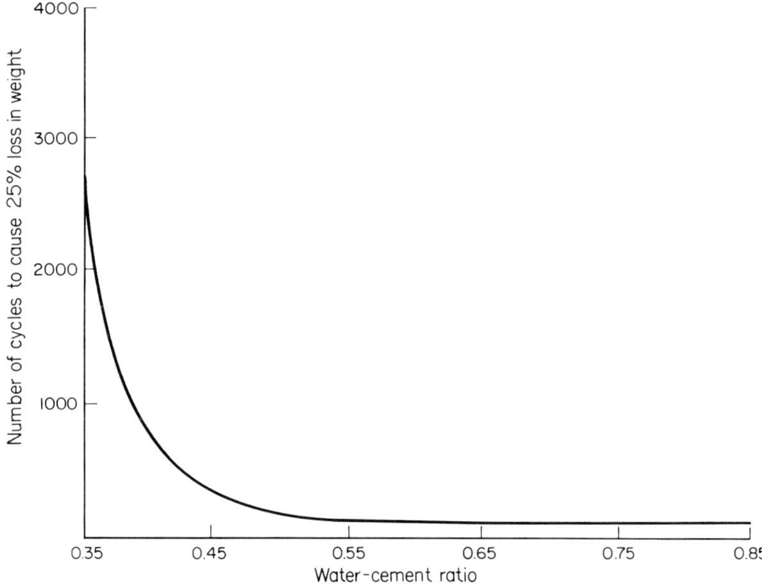

Fig. 1.57 The relationship between water–cement ratio and freeze–thaw resistance.

of the lignosulphonates to provide an improved resistance to freeze–thaw was particularly important in the pozzolan-containing concretes because the substitution of cement by pozzolan generally in this work resulted in a reduction in the freeze–thaw resistance. The use of water-reducing admixtures permitted the inclusion of pozzolan for other beneficial side effects such as reduction of sulphate attack or to enhance resistance to marine environments. Indeed where this latter environment, i.e. sea water, is important, freeze–thaw cycling of concretes with or without sulphite liquor (a crude lignosulphonate) has been carried out [119] in sea water of 34 g l^{-1} salt with the graphical results shown in Fig. 1.58. It can be seen that there is a considerable enhancement in the durability to such conditions in the presence of the lignosulphonate-containing material.

When lignosulphonates are used to reduce the cement content whilst maintaining the workability and strength characteristics, it has been found [111] that there is still a considerable enhancement of durability of those mixes containing less cement and a lignosulphonate plasticizer, in comparison to a control. This is illustrated in Fig. 1.59 where progressive reductions in cement content have been made using a lignosulphonate plasticizer to maintain the 28-day strength.

The recorded data on lignosulphonate water-reducing agents indicate

Table 1.25 Freeze–thaw resistance of concretes used in dam construction

Mix no.	Water–cement ratio	Agent		†Pozzolan (%)	Cycles of freezing and thawing to 25% weight loss	
		%	Type		28 day fog cure	14 day fog plus 76 day 50% r.h.
AINSWORTH CANAL						
59	0.51	0		0	570	780
60	0.53	0.2	G	0	750	720
61	0.52	0.4	G	0	930	850
MONTICELLO DAM						
74	0.57	0.2	B	30	620	580
75	0.50			0	1180	2900
76	0.53			30	650	530
77	0.44	0.1	A	0	1540	2830
78	0.50	0.1	A	30	960	620
79	0.43	0.1	A	30	1490	2050
KIRWIN DAM						
80	0.58	0		0	270	
81	0.52	0.25	L	0	430	1050
82	0.50	0.20	A	0	350	490
GROSS DAM						
83	0.66	0		0	210	1070
84	0.62	0.2	A	0	430	1640
FLAMING GORGE DAM						
96	0.54	0		33.3	630	410
97	0.50	0.37	G	33.3	670	440

Table 1.25 (continued)

Mix no.	Water–cement ratio	Agent		†Pozzolan (%)	Cycles of freezing and thawing to 25% weight loss	
		%	Type		28 day fog cure	14 day fog plus 76 day 50% r.h.
GLEN CANYON DAM						
98	0.51	0		42.8	550	320
99	0.46	0.37	G	42.8	1020	410
100	0.56	0		33.3	800	400
101	0.49	0.37	G	33.3	860	540
104	0.64	0		0	600	900
105	0.60	0.27	G	0	660	800
106	0.56	0.54	G	0	450	450
107	0.66	0		20	530	590
108	0.63	0.27	G	20	530	710
109	0.61	0.54	G	20	590	810
110	0.67	0		20	360	230
111	0.61	0.37	G	20	460	250

†Fly ash used as pozzolan.
Agents A: ammonium lignosulphonate
B: ammonium lignosulphonate
G: calcium lignosulphonate
L: calcium lignosulphonate

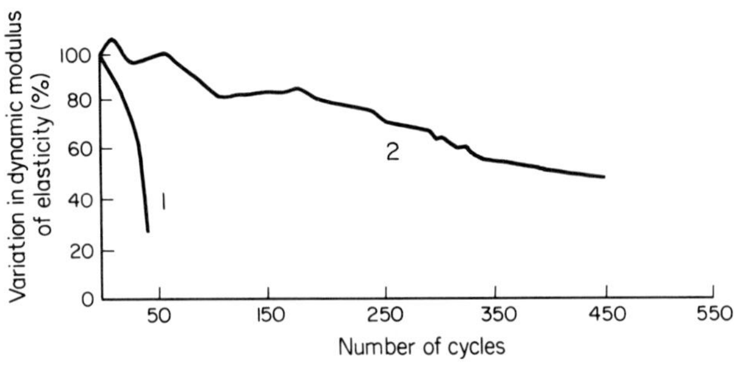

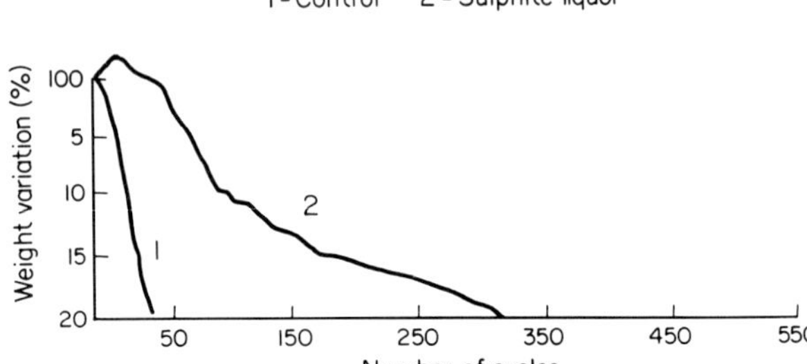

Fig. 1.58 The freeze–thaw resistance of concrete containing lignosulphonates (sulphite liquor) under saline conditions.

that, as far as freeze–thaw durability is concerned, because of the low water–cement ratios possible, an enhancement to the durability will invariably be obtained. When the admixtures are used to effect a reduction in the cement content, there are strong indications that a considerable enhancement of durability is obtained, presumably due to a reduction in the cement matrix which is the part of the concrete susceptible to frost damage. The higher aggregate content would therefore allow easier dissipation of stresses.

(ii) *Hydroxycarboxylic acid based materials.* There is little information on the effect of this class of material on the freeze–thaw durability of concrete into which the admixture is incorporated. However, one set of data is shown in Table 1.26 [120].

(iii) *Air entraining water-reducing admixtures.* Mixes containing water-reducing admixtures based on both lignosulphonate and hydroxycarboxylic acids have been incorporated into air-entrained concrete and compared for freeze–thaw durability in comparison to straight air-entrained mixes. All concretes were designed to have between 5 and 7% of air by volume in the plastic state and the freeze–thaw cycling was carried out under water using a

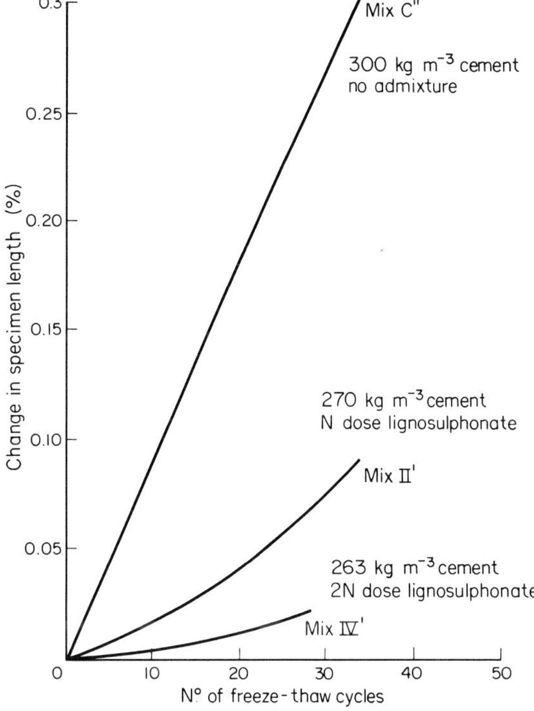

Fig. 1.59 The freeze–thaw resistance of concretes of different cement contents in the presence of lignosulphonates (Hewlett).

cabinet complying with the requirements of ASTM C.29061T. Beams were tested up to 300 cycles of freezing and changes in fundamental longitudinal frequency and weight loss were determined. The changes in frequency were used to determine the durability factor and weight loss was used as a measure of surface deterioration [121]. The results shown in Table 1.27 were obtained.

Table 1.26 Freeze–thaw resistance of concretes containing a hydroxycarboxylic acid water-reducing agent

Admixture	Dose (%)	Cement content of mix (gravel aggregate, slump = 75 to 100 mm) (kg m^{-3})	W/c ratio	Air content	Relative durability to freeze–thaw cycling – number of cycles to reduce dynamic modulus by 50%
None	—	313	0.61	2.7	26
Hydroxycarboxylic acid	0.16	308	0.59	3.5	56

Table 1.27 Freeze–thaw resistance of air-trained concrete containing water-reducing admixtures

Water-reducing admixture	Dosage	Air-entraining agent type	W/c ratio	Freeze–thaw data: Durability factor (%)	Freeze–thaw data: Weight loss (%)
None (control)	—	Vinsol resin	0.53	86	0.48
None (control)	—	Vinsol resin	0.53	84	1.27
(A) Salt of hydroxy-carboxylic acid	1 × normal	Vinsol resin	0.50	99	1.79
	2 × normal	Vinsol resin	0.48	98	0.52
	4 × normal	Vinsol resin	0.45	97	2.05
(B) Calcium lignosulphonate	1 × normal	Vinsol resin	0.47	95	0.92
	2 × normal	Vinsol resin	0.42	95	1.60
	4 × normal	Vinsol resin	0.39	93	0.88
(C) Calcium lignosulphonate	1 × normal	Vinsol resin	0.50	92	0.59
	2 × normal	Vinsol resin	0.47	96	1.64
	4 × normal	Vinsol resin	0.45	97	0.20

It can be seen that the durability to freeze–thaw cycling measured by the durability factor, is enhanced by the reduction in water–cement ratio due to the presence of the water-reducing admixtures of both types. The means of measuring this durability factor, i.e. the change in resonance frequency, would indicate that the internal integrity of the material is maintained at a higher level in the presence of the water-reducing admixtures. However, the weight loss from the surface data are somewhat inconclusive and it does appear that no similar enhancement of resistance to surface spalling is obtained. More recently, freeze–thaw data have been published [122] comparing concrete containing no admixture with concrete air-entrained to the Department of the Environment paving quality concrete specification using neutralized wood resins and an air-entraining water-reducing agent based on hydroxycarboxylic acid. The results are shown in Table 1.28 where

Table 1.28 Freeze–thaw resistance of concrete containing an air-training water-reducing admixture

Mix no.	1	2	3
Concrete containing	no admixture	neutralized wood resin	air-entraining water reducing agent
Air content (%)	1.2	4.5 ± 0.3	4.5 ± 0.3
Freeze–thaw dilation test (% dilation at 50 cycles)	0.035	−0.033	−0.031
DHO test (number of cycles to cause 0.5 mg mm^{-2} scaling)	7.5	15.0	25.0

Table 1.29 Freeze–thaw resistance of corresponding mixes containing water-reducing admixtures and air-entraining agents (after Mielenz)

Series	Mix	Air-entraining agent	Water-reducing admixture	Cement content (kg m^{-2})	W/c ratio	28-day compressive strength (N mm^{-2})	Air content (%)	Slump (mm)	Durability factor ASTM C290 1967
I	F	Vinsol resin	None	261	0.62	28.1	5.6	102	28
	G		Calcium lignosulphonate	227	0.63	26.8	5.2	89	25
II	F	Vinsol resin	None	314	0.53	33.8	5.1	159	34
	G		Calcium lignosulphonate	311	0.45	36.7	6.0	165	51
III	F	Vinsol resin	None	307	0.53	32.9	5.8	159	34
	G		Calcium lignosulphonate	307	0.47	35.8	5.6	171	42

the freeze–thaw dilation tests according to the draft British Standard using temperature cycling under water and also a Canadian test using cycling under salt water and measuring surface scaling, indicate improvements over both the control concrete and the straight air-entrained concrete.

In the area of air-entraining water-reducing admixtures, it is difficult to find data relevent to corresponding mixes where properties of concrete of similar workability and strength characteristics subjected to freeze–thaw cycling are compared, but the data given in Table 1.29 [123] show that when the strength is at least similar to that of the control containing more cement but no admixture, there appears to be no detrimental effect on the durability to freeze–thaw cycling.

(iv) *Superplasticizers.* An increasing amount of data is available on the freeze–thaw testing of concrete containing superplasticizers and Table 1.30 presents results obtained on concrete containing a superplasticizer based on the sodium salt of polymerized naphthalene sulphonic acid both in the form of very workable flowing concrete and at normal workability achieved by considerable reductions in the water–cement ratio. Also shown are results for absorption and initial surface absorption of concrete specimens. This information suggests that there is no deleterious effect on the concrete by the inclusion of the superplasticizer for the production of flowing concrete, and that there is some indication that more durable concrete is produced when the water–cement ratio is reduced. This is true when freeze–thaw cycling under water or under salt solution [33]. Similar results [91] have been obtained for concrete containing the melamine formaldehyde sulphonates. A North American study of melamine formaldehyde sulphonate, naphthalene formaldehyde sulphonate and modified ligno-sulphonate based superplastizers in air-entrained concrete at constant water–cement ratios indicated that the freeze–thaw durability of the superplasticizers was in all cases at least equal to the control air-entrained concrete as shown in Table 1.31 [124].

It can be concluded from the assessment of the data in this section, that inclusion into a concrete mix of a water-reducing admixture of the lignosulphonate, hydroxycarboxylic acid, air-entraining or superplasticizer type should not lead to any deterioration in the durability of that concrete to freeze–thaw cycling. Indeed there are strong indications that, when used either as a means of reducing the water–cement ratio or, alternatively, of reducing the cement content, more durable concrete may result.

(c) Protection of steel reinforcement

One of the prime functions of concrete in load-bearing structures is to protect the steel incorporated to increase the tensile strength of the material.

Table 1.30 Freeze–thaw resistance of superplasticized concrete (Rixom)

Mix No.	1	2	3
Description	Control	Addition of admixture normal workability	Addition of admixture self-compacting concrete
Mix design:			
10 mm gravel (rounded irregular) (kg)	8	8	8
Zone 3 sand (kg)	4	4	4
OPC (kg)	2	2	2
Water (litre)	1.1	0.9	1.0
Admixture (% by weight) of cement)	0	2.5	2.5
Properties of plastic concrete:			
Slump (mm)	60	60	Collapse
Air content (%)	1.8	1.7	2.2
Compacting factor	0.87	0.90	—
Density (kg m^{-3})	2394	2430	2394
Properties of hardened concrete (N mm^{-2}):			
1 day	6.3	12.9	6.0
7 days	31.2	40.2	32.4
28 days	41.2	64.3	42.0
Absorption (% of dry cube weight)			
10 min	1.3	0.8	1.1
30 min	1.9	1.0	2.0
60 min	2.9	1.8	2.8
24 h	5.2	2.6	5.0
ISAT method			
Initial surface absorption test			
BS 1881, Part 5, 1970 Permeability (ml m^2 s^{-1})			
10 min	0.51	0.15	0.40
30 min	0.42	0.08	0.31
60 min	0.26	0.03	0.20
Freeze–thaw dilation test (dilation at 50 cycles, %)	0.10	0.030	0.075
DHO test (number of cycles to cause 0.5 mg mm^{-2} scaling)	6	32	12

If exposed to air and water the steel will rust (an electrochemical reaction), the reaction products having a lower density than the original steel. The expansive forces caused result in stresses being applied internally so that the concrete fails in tension causing large pieces to be broken away at the surface. In view of this, water–cement ratios are kept as low as possible to prevent the ingress of water, and minimum cover of concrete over the reinforcement is often specified.

Table 1.31 The effect of three different types of superplasticizers on the freeze–thaw resistance of air-entrained concrete (Malhotra)

Mix no.	Type of superplasticizer and dosage in % by weight of cement		Mix proportions: Water ($kg\ m^{-3}$)	Cement ($kg\ m^{-3}$)	W/c ratio	Properties of fresh concrete: Slump, (mm)	Air content (%)	Air void system: A (%)	α (cm^{-1})	$\bar{L}$ (μm)	Durability factor %: DF	RDF
1	Control, without AEA		169	402	0.42	45	2.1	—	—	—	—	—
2	Control, with AEA		170	405	0.42	45	4.8	4.9	340	150	100	—
3	Melamine based	1%	158	376	0.42	100	5.2	5.8	246	180		100
4		2%	157	374	0.42	230	5.2	4.3	209	250		100
5		3%	158	376	0.42	260	4.8	3.5	346	150		100
6	Naphthalene based	0.5%	159	379	0.42	90	5.0	4.4	243	200		> 100
7		1.0%	158	376	0.42	260	4.8	4.1	226	230		100
8		1.5%	158	376	0.42	7260	3.4	2.8	308	200		> 100
9	Ligno-sulphonate based	1%	158	376	0.42	100	6.0	4.2	362	150		> 100
10		2%	158	376	0.42	210	6.8	7.1	234	180		> 100
11		3%	158	376	0.42	260	6.0	5.1	310	150		100

In order to study the effect that water-reducing admixtures may have on the role that concrete plays in protecting steel reinforcement, it is necessary to consider the following aspects:

(i) *Concrete permeability*. This has already been dealt with in previous parts of this section where it can be seen that a reduction in the water–cement ratio by use of admixtures is beneficial in reducing the permeability of the concrete. Even when cement contents are reduced, whilst maintaining the workability and strength characteristics of corresponding mixes containing water-reducing admixtures, there is no deleterious effect on the permeability.

(ii) *Chemical attack on the reinforcement.* The action of an admixture in relation to attack on reinforcement can be considered either in direct chemical reaction with the steel or, alternatively, a breakdown of the passive layer imparted by concrete which normally prevents corrosion at the cement/steel interface. In this respect, any accelerating water-reducing

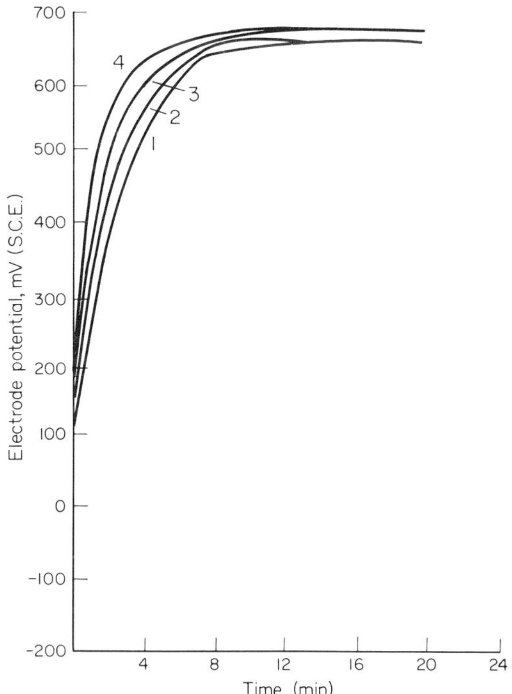

Fig. 1.60 Potential–time curve for steel embedded in cement paste containing black lye. Curve 1 – 0.0% black lye; curve 2 – 0.2%; curve 3 – 0.5%; curve 4 – 1%.

admixtures containing calcium chloride can be considered hazardous as far as raising susceptibility of steel reinforcement to corrosion is concerned. It is particularly so at calcium chloride contents in the concrete at or above 1.5% by weight of cement as discussed in the section on accelerators. The use of such materials has recently been prevented by relevant codes of practice where embedded metal is present in the concrete.

Some data are available [125, 126] concerning lignosulphonate based materials, and Figs 1.60 and 1.61 show, respectively, the potential–time curves for steel embedded in cement paste containing sulphite black lye (an impure lignosulphonate) at various levels, and potential–time curves for steel embedded in cement paste containing 2% calcium chloride and different concentrations of black lye. It is clear that the lignosulphonate-containing admixture not only enhances the passivation of the embedded steel, but also can counteract the effect of calcium chloride on it.

More direct information is available for a superplasticizer based on the formaldehyde polymer of sodium naphthalene sulphonate [127] where the

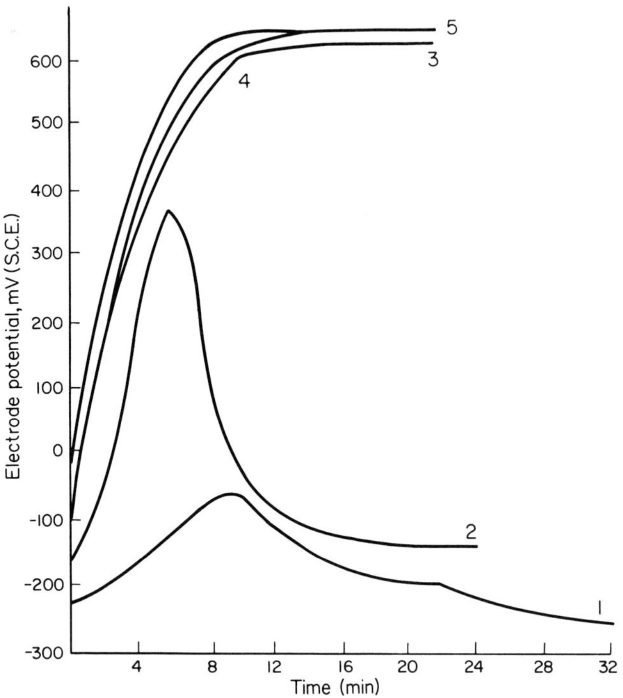

Fig. 1.61 Potential–time curves for steel embedded in cement paste containing 2% $CaCl_2$ and different concentrations of black lye. Curve 1 – 0.0% black lye; curve 2 – 0.2%; curve 3 – 1.5%; curve 4 – 2.0%; curve 5 – 3.0%.

reinforcement was studied after being embedded for 5 years in centrifuged concrete piles. Results were compared with a control concrete containing no admixture and also with concretes containing 1 and 2% by weight of calcium chloride based on the cement content of the concrete. The results are shown in Table 1.32. These data indicate that there is little difference between a control concrete and that containing a superplasticizer, whereas the piles containing calcium chloride show considerable reinforcement corrosion. This concrete is believed to have been steam cured which would tend to accelerate any corrosion process and explains the much greater corrosion than would normally be expected from 1 to 2% $CaCl_2$ in well-compacted concrete.

Table 1.32 Rusting of reinforcement embedded for five years in concrete piles containing superplasticizers and calcium chloride

Admixture	Rust coverage (%) Sample 1	Sample 2	Sample 3	Sample 4	Average
None (control)	0	0	1.6	0.2	0.4
Superplasticizer	Trace	0.1	0.1	Trace	Trace
Calcium chloride 1%	15.2	6.4	10.8	14.4	11.7
Calcium chloride 2%	16.2	17.6	22.6	22.2	19.6

(iii) *Reinforcement bond.* The bond between the concrete and reinforcement is an important engineering aspect of the composite material and any slippage could possibly lead to an easier ingress of air and moisture leading to corrosion. Information has been obtained on the effect that calcium lignosulphonate has on the bond of concrete with steel and, in one particular investigation [125], the effect before and after the application of anodic-connected direct current. These results are shown in Table 1.33 where it can be seen that the presence of calcium lignosulphonate increases the bond strength in comparison to the control over the applied voltage range.

Similar data have been obtained for other lignosulphonate based water-reducing admixtures where cement and workability were kept constant with a reduction in the water–cement ratio and the reinforcement bond was

Table 1.33 Effect of a lignosulphonate based water-reducing agent on reinforcement bond under an applied external voltage (after Kondo)

	Applied voltage (V) 0		10		20	
Admixture	No	Yes	No	Yes	No	Yes
Bond (N mm^{-2})	3.2	3.4	3.3	3.5	3.7	3.8

Table 1.34 Effect of a lignosulphonate based water-reducing agent on reinforcing bond

Admixture	W/c ratio	Reinforcement bond Horizontal	Vertical
None	0.62	100	100
Lignosulphonate	0.52	121	116

measured by ASTM C234: 1950 Part 4, with the result shown in Table 1.34 [127].

The steel–concrete bond strength of superplasticized concrete in both normal and lightweight concrete has been reported [124] where the superplasticizer was used to increase the workability of the concrete. The bond strengths obtained for smooth and twisted bars are shown in Table 1.35 where the considerable improvements are clearly demonstrated.

The data presented in this section illustrate that, with the exception of those accelerating water-reducing admixtures containing calcium chloride, there is an abundance of evidence to support the conclusion that water-reducing admixtures of lignosulphonate and superplasticizers of the sodium naphthalene sulphonate chemical form certainly will not accelerate any kind of corrosion with reinforcement and, when used to reduce the water–cement ratio, will form a more permeable and durable protective cover for the reinforcement. In view of the chemical nature of other types of materials such as the hydroxycarboxylic acids and the melamine formaldehyde sulphonates, it seems most likely that these materials too would have no deleterious effect in this respect.

(d) Gain in compressive strength

As has been clearly shown in the preceding section, it is possible to produce concrete containing an admixture having a 28-day strength equal to a control concrete but being designed at a significantly lower cement content. However, the choice of 28-day compressive strengths is purely arbitrary and, as far as durability is concerned, it is important that the concrete continues to gain in strength over subsequent years. There are some data available in this area where concretes containing various types of water-reducing admixtures have been tested for strength at ages up to 13 years and Fig. 1.62 shows a graphical summary of this information. The filled points indicates the type of results obtained for up to 50 years on many hundreds of concrete specimens not containing an admixture and it can be seen that by and large the available data fit in the filled points at least as often as results for tests on control concretes carried out during the admixture evaluations.

In view of this it seems reasonable to conclude that of the five types of water-reducing admixture reported, there is no indication of any subsequent loss of strength or of any significant decrease in the gain of strength.

Table 1.35 Effect of superplasticized concrete on the steel-concrete bond (Malhotra)

Type of concrete	Type of admixture used	Cement brand and content		Slump	Bond strength, kg/m^2			
					7-day		28-day	
		Cement brand	Cement content		Smooth bars	Twisted bars	Smooth bars	Twisted bars
Normal weight	None	Type I, Brand 1	400	100	12	150	13	152
Normal weight	Superplasticizer	Type I, Brand 1	400	220	35	275	40	285
Lightweight ($1800 kg/m^3$)	None	Type I, Brand 2	500	100	4	66	6	92
Lightweight ($1800 kg/m^3$)	Superplasticizer	Type I, Brand 2	500	210	9	142	21	210

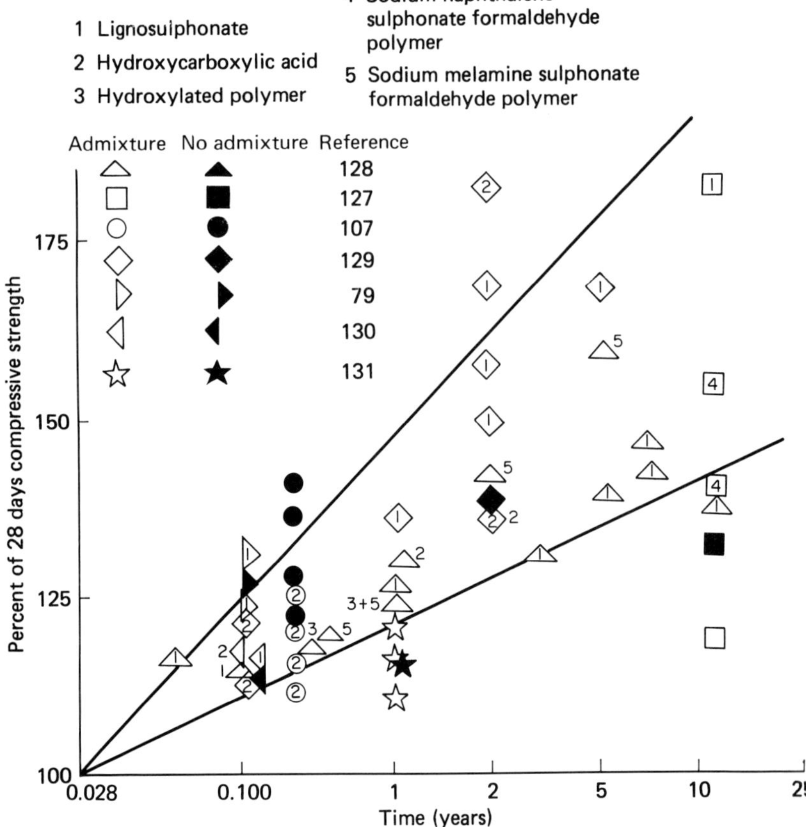

Fig. 1.62 The increase in compressive strength of concretes containing various water-reducing agents.

(e) Volume deformations

The volume deformations of concrete are shrinkage, which occurs under drying conditions, and creep, which is the additional deformation obtained under an applied stress. Creep does occur under saturated conditions (basic creep) but increases considerably under conditions of moisture loss. The picture is rather complicated in that creep is made up of a recoverable and irrecoverable portion on removal of the applied stress.

Both deformations are believed to proceed by the same mechanism which, although not fully understood, is associated with the bound water in the tobermorite gel. These volume deformations are recognized as important design considerations and are compensated for in codes of practice for structural design. It will be appreciated that any change in the known behaviour of concrete by the addition of minor ingredients could result in durability problems in concrete structures; increased deflection in beams and slabs, the transfer of load to reinforcement, loss of prestress in prestressed concrete, changes in relative movements between components, and changes in stability of slender columns and walls [132] could all lead to minor or major failures.

Volume deformations are largely a function of the nature and quantity of the cement paste in the concrete and it has been shown [133] that studies on cement paste properties correlate well with those of the concrete into which it is incorporated, certainly as far as comparison of the effect of added materials.

It is difficult to prepare a definitive synopsis of the effect that the various classes of admixtures will on the volume deformations of concrete because of the conflicting results between workers and different types of test methods used, but the points given below should enable some guide lines to be set.

(i) *Volume deformations under saturated conditions.*
(1) *Shrinkage/swelling.* Under saturating conditions where there is no weight loss due to the evaporation of water, there is little volume change in cement paste specimens for admixtures based on lignosulphonate alone, and accelerating water-reducing admixtures based on lignosulphonate with the addition of either calcium chloride or triethanolamine [134] (Fig. 1.63). This has been studied in greater detail [107] for a hydroxycarboxylic acid based material containing no additional accelerating materials where it can be seen in Fig. 1.64 that any shrinkage/swelling properties are not significantly changed by the addition of the admixture and are largely a function of the thermal expansion of the resultant concrete.
(2) *Basic creep.* The creep of concrete under near saturated conditions (98% r.h.) has been studied using a lignosulphonate water-reducing admixture, a lignosulphonate with the addition of calcium chloride, and a ligno-sulphonate with the addition of triethanolamine [135]. The mix designs used in this work were identical for each mix, therefore the water-reducing admixtures were used as direct additions to increase the workability. The results are shown in Table 1.36 where it can be seen that there are only marginal increases in creep over the plain mix for each of the admixture types studied.

Statistical evaluation of all the results indicated that, at the 90% confidence limit, there was no significant difference in the creep of any of the mixes.

Some results are available for the hydroxycarboxylic acid based materials [107] and these data are presented graphically in Fig. 1.65.

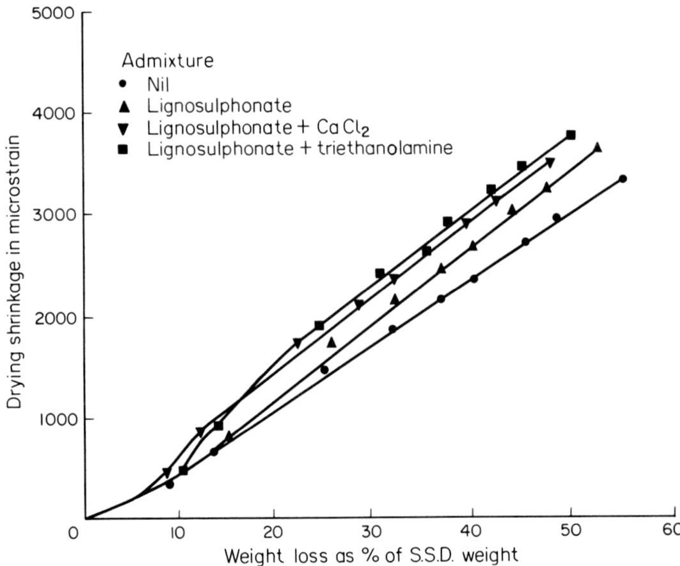

Fig. 1.63 Volume changes of cement pastes under near saturated conditions (Morgan).

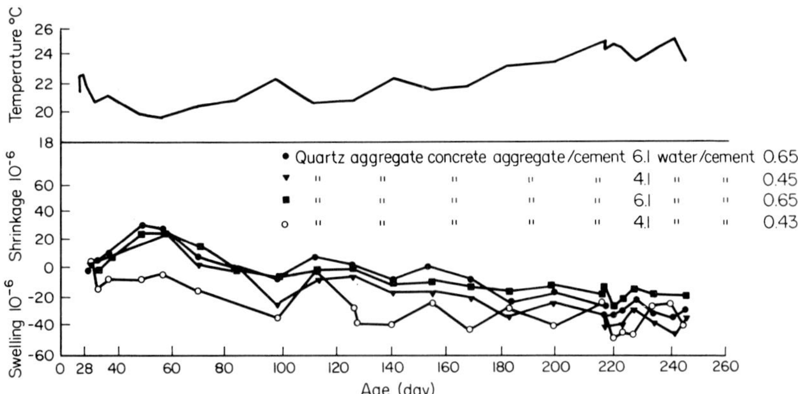

Fig. 1.64 Volume changes of concretes containing a hydroxycarboxylic acid water-reducing agent under saturated conditions (Neville).

Table 1.36 Effect of lignosulphonate based admixtures on basic creep of concrete

Mix details	Admixture	Mix designation	Observed creep microstrain	% increase in creep over plain mix
Direct addition to plain mix 98% r.h. 56 days under load	Nil	AW	240	—
	Lignosuphonate	AW1	270	+ 12
	Lignosulphonate + $CaCl_2$	AW2	250	+ 4
	Lignosulphonate + triethanolamine	AW3	270	+ 12

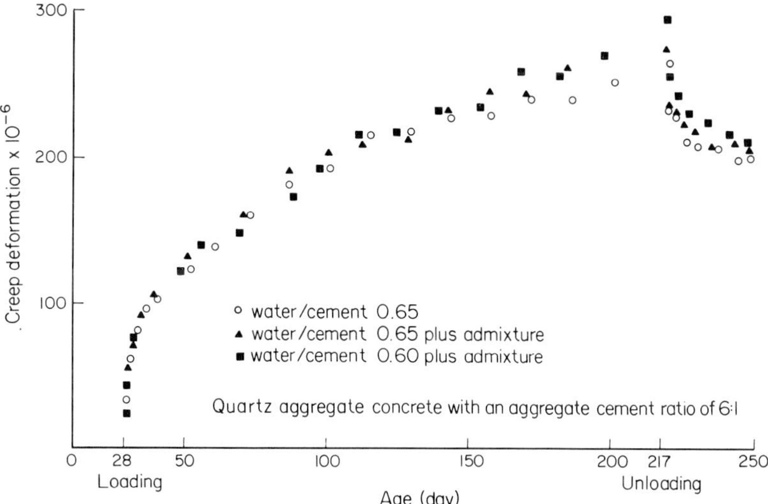

Fig. 1.65 The creep and creep recovery of concrete containing a hydroxycarboxylic acid water-reducing agent under saturated conditions (Neville).

(ii) *Volume deformations under drying conditions.* It is under conditions where moisture is lost from concrete that volume deformations under loaded or unloaded conditions occur to any magnitude. It is difficult to say what degree of relative humidity structural concrete will be subjected to in actual practice, but certainly for thin sections or near the surface of large sections, considerable interchange of water due to changing climatic conditions will occur.

Although there are some anomalies in the literature, it is generally agreed that both types of volume deformation are a function of the same fundamental mechanism and that the influence of other factors such as admixtures will affect both shrinkage and creep in a similar manner. As outlined earlier, water-reducing admixtures can be used to obtain different effects on the plastic/hardened concrete and it is this factor, together with the admixture type, that is important in determining the effect on the volume deformations of concrete.

(1) *Direct addition of water-reducing admixture* to increase the workability of the concrete. The effect of all types of water-reducing admixture under these conditions is invariably to increase the shrinkage and creep of the concrete. Some typical values are shown in Table 1.37.

The considerable increases in both shrinkage and creep in the presence of admixtures containing calcium chloride and triethanolamine are clearly illustrated.

Table 1.37 Effect of the direct addition of water-reducing admixtures on the drying creep and shrinkage of concrete

Type of water-reducing admixture	% r.h.	% increase over plain mix: Shrinkage	Creep	Reference
Normal lignosulphonate	50	10* 1†	34‡	[19, 135]
	50	15§	15§	[79]
Hydroxycarboxylic acid	50	(−4) to (2)		[136]
Accelerating lignosulphonate + $CaCl_2$	50	19* 4†	54‡	[19, 135]
Accelerating lignosulphonate + triethanolamine	50	35* 13†	79‡	[19, 135]

* After 14 days drying.
† After 203 days drying.
‡ After 84 days drying.
§ After 175 days drying.

(2) *Admixture addition with simultaneous reduction in cement content* to produce corresponding mixes having similar 28-day strengths and workability characteristics to a plain concrete control, has been studied with regard to volume deformations and, although again there are some conflicting results, it does seem that lignosulphonate materials used in this context produce little if any change in the creep and shrinkage characteristics of the concrete. However, where triethanolamine and calcium chloride have been included in the formulations to produce accelerating materials, it is possible to obtain increases in the creep and shrinkage, and a limited amount of work on hydroxycarboxylic acids suggest that even in corresponding mixes there may be an increase in volume deformations. (See Table 1.38). It will be appreciated from the spread of results and the fact that many tests were carried out under different conditions, that it is difficult to make comparisons, and only the trends can be observed.

Table 1.38 Effect of water-reducing admixtures on the drying creep and shrinkage of corresponding mixes

Type of water-reducing admixture	% r.h.	% increase over plain mix: Shrinkage	Creep	Reference
Normal lignosulphonate	50	− 4 to + 22	− 3 to + 17	[79]
	35	—	+ 2 to + 17	[137]
	95	—	− 2 to + 19	[130]
	50	+ 1	—	[19]
	50	+ 1 to + 4	—	[136]
Retarding hydroxycarboxylic acid	95	—	− 32 to + 13	[130]
	94	− 52 to − 6	− 5.8 to + 49	[138]
Accelerating	50	+ 3	—	[19]
lignosulphonate + $CaCl_2$	50	+ 11	—	[19]
Accelerating lignosulphonate + TEA	35	—	− 8 to + 35	[137]

(3) *Addition of admixture to obtain higher strengths.* In this situation the small amount of reported work indicates that where the water–cement ratio is reduced, the shrinkage and creep of the concrete is also reduced. Fig. 1.66 shows the effect of a hydroxycarboxylic acid plasticizer on the creep of the concrete where the material has been used to effect a reduction in the water–cement ratio without any other changes in the mix design. Thus it seems that the reduction of the water–cement ratio will compensate for the increases in creep observed in the data above.

This is a difficult field for interpretation where techniques and procedures vary from worker to worker and where conflicting results are apparent.

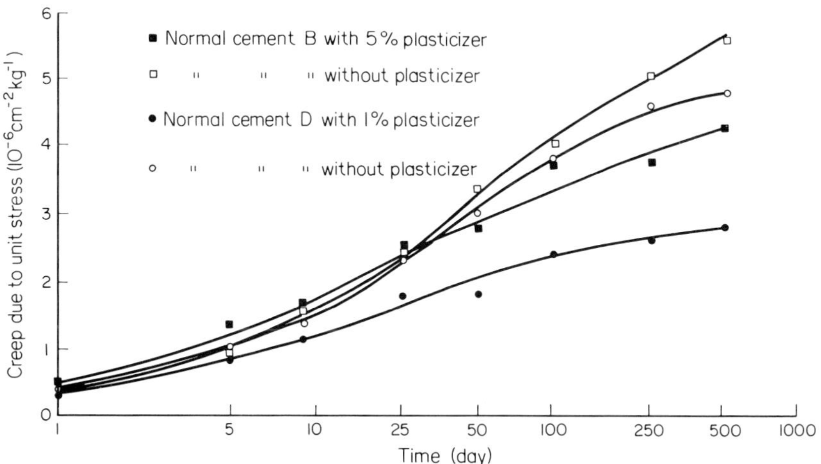

Fig. 1.66 The effect of a hydroxycarboxylic acid water-reducing agent on the creep of concrete when used to lower the water–cement ratio (Rodrigues).

However, it is felt that the intrinsic property of most water-reducing admixtures is to increase the creep and shrinkage of the concretes into which they are incorporated. This is minimized by utilizing the effect of the admixture to reduce the cement and, therefore, the paste content of the mix. For corresponding mixes with lignosulphonate water-reducing agents, the current FIP/CEB recommendation for creep computation can be used with confidence because of the incorporation of a cement content/water–cement ratio function in the calculation.

When materials other than a lignosulphonate water-reducing agent are used, in conditions where significant drying of concrete can take place, caution must be exercised where straight additions are used and where differential creep rates could be obtained in comparison to adjacent concrete containing no admixtures.

It is interesting to speculate on the reasons for such varied effect on creep and it is felt that although current theories have been based on the effect on the tobermorite gel layering [139, 140] some thought should be given to the possibility that it is the known interference (see earlier) on the ettringite reaction of water-reducing admixtures. Indeed the sulphate component of cement has been shown to influence the shrinkage of cement paste [133] according to the following equation:

$$\text{paste shrinkage (microstrain)} = (120 + 260\ C_3A + 101\ C_4AF - 770\ SO_4 + 1200\ \text{alkali} + 0.4\ \text{fineness}).$$

It can be seen, therefore, that on typical paste shrinkage in the region of 4000 microstrain, that some 30 to 40% of this could be accounted for by the sulphate component. An important point to note is that the sulphate reaction has a negative influence on shrinkage and, therefore, acts as a restraint to creep and shrinkage as the reaction proceeds. It was noted in an earlier section that the addition of a water-reducing admixture delays ettringite reactions and could be a possible mechanism by which the volume deformations are increased.

1.6.3 Durability guidelines

(i) Lignosulphonate admixtures can be used to produce concrete of a required workability and strength characteristic at lower cement contents than the comparative plain concrete with no adverse effect on the durability of the concrete or total structure. The only exception to this rule would be in conditions where high sulphate ground waters may be involved when the minimum cement contents of relevant codes of practice should be observed.

(ii) Water-reducing admixtures containing calcium chloride should not be used in concrete containing embedded metal or where volume deformations are important.

(iii) In shrinkage and creep sensitive situations, sections of a concrete structure should not be treated with any type of water-reducing admixture either as a means of obtaining extra workability or for the retempering of slightly matured mixes where the differential volume changes of adjacent parts of the structure could lead to cracking. This is also true, of course, when cement contents are increased for the same reasons.

(iv) Lignosulphonate admixtures containing triethanolamine should not be used in situations sensitive to increased volume deformations.

References

1 US Patent (1932). 643 740.
2 Ros, M. (1934). *Eigenschoften des Betons,* Report No. 79. EMPA Zurich.
3 US Patent (1937). 2081 642.
4 US Patent (1937). 2081 643.
5 US Patent (1938). 2127 451.
6 US Patent (1938). 2141 570.
7 US Patent (1939). 2169 980.
8 US Patent (1939). 2174 051.
9 US Patent (1941). 2229 311.
0 Cement Admixtures Association/Cement and Concrete Association (1975). *Superplastics for Concrete,* p. 88.
11 Cook, H.K. (1967). *Proceedings of the International Symposium on Admixtures for Mortar and Concrete,* Brussels, 135–6.
12 Mouton, Y. (1972). *Bulletin of the Liason Laboratory,* Ch. 58, Ref. 1185.
13 Edmeades, R.M. and Hewlett, P.C. (1975). *Proceedings of the First International Congress on Polymer Concretes.* London, Section 7, Paper 10.
14 Hansen, W.C. (1959). *ASTM Special Publication,* **266,** 20–2.
15 Chaiken, B. (1961). *Public Roads,* **31,** 126–35.
16 Joisel, A. (1973). *Physico Chemistry of Admixtures for Cement and Concrete,* 40.
17 Ernsberger, F.M.Z. (1948). *Journal of Physics and Colloid Chemistry,* **52,** 267–76.
18 Rezanowich, A. (1960). *Journal of Colloid Science,* **15,** 452–71.
19 Morgan, D.R. (1974). *Proceedings of the First Australian Conference on Engineering Materials,* University of New South Wales, 97–108.
20 Foster, D.E. (1963). *ACI Journal* Title **60–64,** 1481–523.
21 Hewlett, P.C. (1975). Private communication.
22 Anon (1967). *Admixtures for Concrete.* Concrete Society TRCS, 1.
23 Anon (1974). *Heptonates as additives for Concrete and Cement,* Croda Ltd., data sheet and Sandberg reports L337 and M13.
24 Bruere, G.M. (1964). *Constructional Review, Australia,* **37,** 16–21.
25 Diamond, S. (1971). *Journal of the American Ceramic Society,* **54,** 273–6.
26 Mielenz, R.C. (1968). *Proceedings of the 5th International Symposium of Cement,* Tokyo, 32.
27 Danielson, U. (1967). *Proceedings of the International Symposium on Admixtures for Mortar and Concrete,* Brussels, 58–67.
28 British Patent (1967). 1068 886.
29 British Patent (1973). 1386 933.
30 US Patent (1970). 3537 869.

31 Japanese Patent (1972). 50329/72.
32 Japanese Patent (1973). 53192/73.
33 Rixom, M.R. (1974). *Precast Concrete Journal,* **5,** 633–7.
34 Canadian Patent (1975). 961 866.
35 Karsten, R. (1967). *Proceedings of the International Symposium on Admixtures for Mortar and Concrete,* Brussels, 357.
36 Rixom, M.R. (1975). *Proceedings of the Workshop on the use of Chemical Admixtures in Concrete,* University of New South Wales, 153.
37 Gilbert, E.E. (1968). *Sulphonation and Related Reactions,* Wiley Interscience, New York.
38 US Patent (1956). 2730 516.
39 British Patent (1969). 1169 582.
40 Aignesberger, A. (1943). *Cement, Lime and Gravel,* **48,** 188–92.
41 Davis, B. (1975). *Proceedings of the First International Congress on Polymer Concretes,* London, 6.
42 Japanese Patent (1974). 80 133.
43 Japanese Patent (1972). 41 897.
44 Japanese Patent (1975). 36 517.
45 Japanese Patent (1973). 117 519.
46 Japanese Patent (1974). 71 416.
47 Hanson, W.C. (1972). *Journal of Materials JMSLA,* **5,** 842–55.
48 Ernsberger, F.M. (1945). *Industrial and Engineering Chemistry,* **37,** 598–600.
49 Manabe, T. (1959). *Japan Cement Engineering Association, Review,* 40–6.
50 Blank, B. *et al.* (1963). *Journal of the American Ceramics Society,* **46,** 395–9.
51 Rossington, D.R. (1968). *Journal of the American Ceramics Society,* **51,** 46–50.
52 Mielenz, R.C. (1968). *Proceedings of the Fifth International Symposium on the Chemistry of Cement,* Tokyo, 15–19.
53 Ramachandran, V.S. (1971). *Cement Technology,* **2,** 121–9.
54 Ramachandran, V.S. (1972). *Cement and Concrete Research,* **2,** 179–94.
55 Young, J.F. (1972). *Cement and Concrete Research,* **2,** 415–33.
56 Flateau, A.S. (1974). *Concrete,* **8,** 45–7.
57 Diamond, S. (1972). *Journal of the American Ceramics Society,* **55,** 405–8.
58 Dierkes, P. (1973). *Proceedings of the Melment Symposium,* Trostberg, 5–24.
59 Zhuravhev, U.F. (1952). *Journal of Applied Chemistry (USSR),* **25,** 1317–24.
60 Prior, M.E. (1959). *ASTM Special Publication No. 266,* 173–4.
61 Singh, W.B. (1975). *Cement and Concrete Research,* **5,** 545–50.
62 Roberts, M.H. (1967). *Proceedings of the International Symposium on Admixtures for Mortar and Concrete,* Brussels, 7–27.
63 Dodson, V.H. (1967). *Proceedings of the International Symposium on Admixtures for Mortar and Concrete,* Brussels, 59–64.
64 Taylor, H.F.W. (1966). *The Chemistry of Cements,* RIC Lecture Series, No. 2.
65 Marianpolski, N.A. (1974). *Neft. Khoz,* **10,** 27–30.
66 Stein, H.W. (1961). *Journal of Applied Chemistry,* **42,** 474–82.
67 Forrester, J. (1967). Private communication.
68 Singh, W.B. (1975). *Cement and Concrete Research,* **5,** 548.
69 Chatterji, S. (1967). *Indian Concrete Journal,* **4,** 151–60.
70 Khalil, S.M. (1973). *Cement and Concrete Research,* **3,** 677–88.
71 Ciach, T.D. (1971). *Cement and Concrete Research,* **1,** 159–76.
72 Ghosh R.S. and Malhotra, V.M. (1978). *Use of Superplasticizers as Water Reducers,* CANMET Division Report MRP/MRL 78–189 (J). CANMET, Energy, Mines and Resources Canada.

73 Edmeades, R.M. (1975). *Proceedings of the Conference on Ready-Mixed Concrete,* Dundee, Theme 3.
74 Hodkinson, K. (1974). Private communication.
75 Hodkinson, K. (1974). Private communication.
76 Rixom, M.R. (1975). *Chemistry and Industry,* **7,** 162–5.
77 Anon (1967). *Admixtures for Concrete,* Concrete Society, Report 1.
78 Fletcher, K.E. (1971). *Concrete,* **5,** 175–9.
79 Tam, C.T. (1974). *Proceedings of the First Australian Conference on Engineering Materials,* University of New South Wales, 73–96.
80 Warris, B. (1967). *Proceedings of the International Symposium on Admixtures for Mortar and Concrete,* Brussels, 30–3.
81 Roeder, A.R. (1976). Private communication.
82 Sasse, H.R. (1975). *Proceedings of the First International Congress on Polymer Concretes,* London, Session E, Paper 2.
83 Howard, F.L. *et al.* (1959). *ASTM Special Publication No. 266,* 149–54.
84 Ravina, D. (1975). *ACI Journal,* **72,** 291–5.
85 Hersey, A.T. (1975). *ACI Journal,* **72,** 526–8.
86 McIntosh, J.D. Private communication.
87 Anon (1976). *Superplasticising Admixtures in Concrete,* Cement and Concrete Association. p.16.
88 Collopardi, M. Corradi, M. and Valente, M. (1979). *Proceedings TRB Symposium in Superplasticizers, Transportation Research Record No. 720,* Washington, D.C.
89 Mailvaganam, N.P. (1978). *Proceedings of the International Symposium on Superplasticized Concrete,* CANMET ,Energy, Mines and Resources. Canada, 649–72.
90 British Patent, (1975). 1391 818.
91 Van der Zanden, J.P.A.J. (1975). *Cement,* **27,** 148–55.
92 Dodson, V.H. (1976). *ACI Journal,* **73,** 1739–53.
93 Anon (1976). *Superplasticising Admixtures in Concrete,* Cement and Concrete Association, p. 17.
94 Vollick, C. A. (1959). *ASTM Special Publication No. 266*, 194–5.
95 Okada, K. (1967). *Proceedings of the International Symposium on Admixtures for Mortar and Concrete,* Brussels, 115–29.
96 Roeder, A.R. Private communication.
97 Shacklock, B.W. (1975). *Concrete Constituents and Mix Proportions,* Cement and Concrete Association.
98 Anon (1974). *ACI Publication SP46,* 97–108.
99 Cordon, W.A. (1975). *ACI Journal,* **72,** 46–9.
100 Pomeroy, C.D. (1972). *C & CA Technical Report* UDC 666, 972, **12:** 666, 97. 017.
101 Lawrence, C.D. (1974). *Proceedings of the International Symposium on Pore Structure and the Properties of Materials,* RILEM/IUPAC, **5,** 167–76.
102 Kromlos, K. (1970). *Magazine of Concrete Research,* **22,** 232–8.
103 Wallace, G.B. (1959). *ASTM Special Publication No. 266,* 44–7.
104 Kreiger, P.C. (1967). *Proceedings of the International Symposium for Mortar and Concrete,* Brussels, 28–9.
105 Larson, T.D. *et al.* (1963). *ACI Journal,* **60,** 1739–53.
106 Authors' data (1975).
107 Brookes, J.J. (1975). *Concrete,* **9,** 33–5.
108 Blundell, R. (1975). Private communication.
109 Anon (1971). *Supplement to the Counsulting Engineer,* **27,** 9.

110 Levitt, M. (1971). *Journal of non-destructive testing,* **12,** 106–12.
111 Hewlett, P.C. (1975). *Proceedings of the Workshop on the use of Chemical Admixtures in Concrete,* University of New South Wales, 43–73.
112 Anon (1970). *Materiaux et Construction,* **3,** 125.
113 Anon (1970). *Materiaux et Construction,* **3,** 130.
114 Ore, E.L. (1959). *ASTM Special Publication No. 266,* 86.
115 Stolnikov, U.V. (1967). *Proceedings of the International Symposium on Admixtures for Mortar and Concrete,* Brussels, 37.
116 Brooks, J.J., Wainwright, P.J. and Neville, A.M. (1978). *Proceedings of the International Symposium on Superplasticized Concrete,* CANMET, Energy, Mines and Resources, Canada, 425–50.
117 Anon (1960). *Conf. Laboratory Report No. C. 810.* US Bureau of Reclamation.
118 Ore, E.L. (1959). *ASTM Special Publication No. 266,* 84.
119 Batarakov, V.G. (1967). *Proceedings of the International Symposium on Admixtures for Mortar and Concrete,* Brussels, 76.
120 Fischer, H.C. (1959). *ASTM Special Publication No. 266,* 214.
121 Larson, T.D. *et al.* (1963). *ACI Journal,* **60,** 1739–53.
122 Rixom, M.R. (1975). *Proceedings of the Workshop on the use of Chemical Admixtures in Concrete,* University of New South Wales, 149–76.
123 Mielenz, R.C. (1968). *Proceedings of the Fifth International Symposium on the Chemistry of Cement,* Tokyo, 11.
124 Malhotra, V.M. and Malanka, D. (1978). *Proceeding of the International Symposium on Superplasticized Concrete,* CANMET. Energy, Mines and Resources, Canada, 673–708.
125 Kondo, Y. *et al.* (1959). *ACI Journal,* **56,** 299–312.
126 Gouda, U.K. *et al.* (1973). *Journal of Colloid and Interface Science,* **43,** 294–302.
127 Anon (1976). *Superplasticizing Admixtures in Concrete,* Cement and Concrete Association, 28.
128 Browne, R.D. (1976). Private communication.
129 Wallace, J.J. (1959). *ASTM Special Publication No. 266,* 38–96.
130 Hope, B.B. (1970). *ACI Journal,* **67,** 673–8.
131 Kobayashi, M. (1967). *Proceedings of the International Symposium on Admixtures for Mortar and Concrete,* Brussels, 79–96.
132 Browne, R.D. (1971). Private communication.
133 Roper, H. (1974). *Proceedings of the First Australian Conference on Engineering Materials,* University of New South Wales, 45–71.
134 Morgan, D.R. (1974). *Materiaux et Construction,* **7,** 283–9.
135 Morgan, D.R. (1973). Ph.D. Thesis, University of New South Wales.
136 Bruere, G.M. *et al.* (1971). *CSIRO Report PB 203. 263,* Australia.
137 Hope, B.B. (1971). *ACI Journal,* **68,** 361–5.
138 Hope, B.B. (1967). *Proceedings of the International Symposium on Admixtures for Mortar and Concrete,* Brussels, 19–32.
139 Morgan, D.R. (1974). *Civil Engineering Transactions,* **56,** 7–10.
140 Jessop, E.L. *et al.* (1968). *Proceedings of the Fifth International Symposium on the Chemistry of Cement*, Tokyo, Supplement IV V, 36–41.

2
Air-entraining agents

2.1 Background and Definitions

The air-entraining admixtures are organic materials, usually in solution form, which when added to the gauging water of a concrete mix, entrain a controlled quantity of air in uniformly dispersed microscopic bubbles. This type of air should not be confused with 'entrapped air' which is often present in concrete in the form of irregularly shaped cavities and which can be due to inadequate compaction or flaky aggregates.

There are three major reasons for intentionally entraining air into concrete: durability, cohesion, and density.

(a) Durability

During the 1930s it was observed that certain stretches of road in the north-east states of America were more able to withstand the effects of freeze–thaw conditions and the presence of de-icing salts than other roads in the area [1]. An investigation revealed that the more durable roads were less dense and that the cement had been obtained from mills where beef tallow had been used as a grinding aid. It was concluded that the beef tallow had functioned as an air-entraining agent and had enhanced the durability of the concrete. This led to a more controlled investigation, and in 1939 an air-entrained concrete carriageway was intentionally produced by the New York Department of Public Works [2].

The effect of the entrainment of a minor amount (4 to 6% by volume) of air on the ability of concrete to withstand freeze–thaw cycling is remarkable [3] and is illustrated in Fig. 2.1.

Fig. 2.1 The effect of 300 freeze–thaw cycles (according to ASTM C666 Procedure B) on air-entrained (right) and plain concrete (left).

In the UK the air entrainment of concrete carriageways and aircraft runways is now accepted as normal practice and several hundreds of kilometres of this type of concrete pavement have been placed. In North America the use of air-entrained concrete is universally accepted and therefore finds greater use than in Europe.

(b) Cohesion

Concrete which is produced using fine aggregates deficient at the fine end of grading, e.g. sea dredged aggregates, exhibit a tendency to bleed and segregate. The presence of a small amount of entrained air (2 to 4% by volume) leads to an improvement in cohesion, or mix stability. Alternatively, with mixes which are adequate in this respect, a reduction in sand content can be made when air is entrained without loss of cohesion. The amount that can be removed is approximately equal on a volume basis and leads to a reduction in water–cement ratio to minimize the effect of entrained air on compressive strength.

(c) Density

The two applicational areas above are normally associated with minor quantities (less than 8% by volume) of entrained air. However, using different chemical types of materials, much larger quantities (up to 30% by volume) can be entrained to lower the density, enhance the thermal insulation properties, or to produce lightweight concrete in conjunction with lightweight aggregates.

When air entrainment is used in the above broad applications to achieve a given end result, there are a number of side effects which need to be considered: (i) the presence of microscopic air bubbles acts as a 'lubricant' and increases the workability, or allows a consequential reduction in the water–cement ratio; (ii) the compressive and tensile strengths decrease with increasing air content when the mix design is unchanged; (iii) the yield of the concrete is increased for a given weight of mix ingredients. Those aspects will be quantified later.

2.2 The chemistry of air-entraining agents

The literature describes many different chemical surfactants as suitable for the formulation of air-entraining agents for concrete. However, in practice, the major proportion of commercial products are based on a relatively small number of raw materials and these are set out below in order of probable decreasing use.

(a) Abietic and pimeric acid salts
(b) Fatty acid salts
(c) Alkyl-aryl sulphonates
(d) Alkyl sulphates
(e) Phenol ethoxylates.

2.2.1 Abietic and pimeric acids

The formula of these chemicals is shown in Fig. 2.2 although they are rarely, if ever, isolated and used as the pure materials. In fact, they are present as major components in resins extracted from pine stumps, or alternatively, from resins derived from tall oil processing. When obtained from wood resins they are present partly as an ester and are subjected to a saponification/neutralization stage with caustic soda to produce the sodium salts. In the case of resin acids, neutralization with caustic soda is carried out directly. Commercial products may contain 5 to 20% active material as a solution in water.

When the material is to be used as a foaming agent for lightweight concrete, a proportion of animal glue is added to further stabilize the bubble

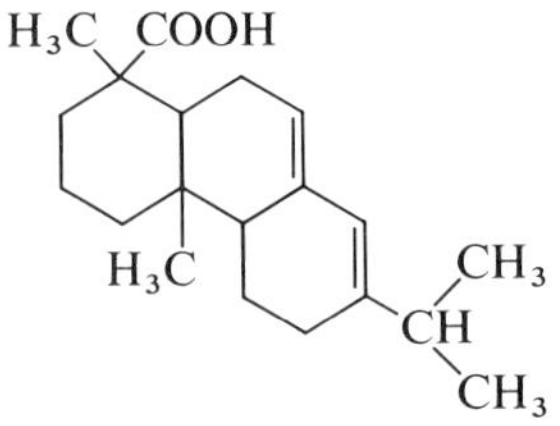

H_3C COOH

H_3C CH_3 CH CH_3

Abietic acid

Fig. 2.2 Abietic and pimeric acids.

structure (see later). A typical preparation [4] is as follows: 4 parts animal glue, 2 parts wood resin, 5 parts 20% caustic soda and 10 parts of water (by weight) are heated to 80 °C and stirred vigorously until a uniform consistency is obtained.

2.2.2 Fatty acid salts

In order to satisfy the requirements of both performance in use and the ability to form stable aqueous solutions of adequate strength, the fatty acids shown in Table 2.1 are used as air-entraining agents for concrete in the form of their alkali metal salts. These fatty acids are present in a distribution of chain lengths in naturally occurring fats and oils, such as tall oil and coconut oil and are used in the form of such a mixture. Typical formulations contain from 5 to 20% by weight of fatty acid salt. These products, unlike the neutralized wood resins, are compatible in solution with certain lignosulphonates and hydroxycarboxylic acid salts to form admixtures possessing both air-entraining and water-reducing capabilities.

Table 2.1 Fatty acids used as air-entraining agents

Fatty acid	Formula	Reference
Oleic acid	$CH_3—(CH_2)_7—CH{=}CH—(CH_2)_7COOH$	[5]
Capric acid	$C_9H_{19}COOH$	[6]

2.2.3 Alkyl-aryl sulphonates

The alkyl-aryl sulphonates tend to find application in the production of lightweight concrete and not in enhancing the freeze–thaw durability of normal concrete. The usual raw material is orthododecylbenzene sulphonate, which is a basic surfactant used in a variety of industrial and domestic detergents. The formula is shown in Fig. 2.3.

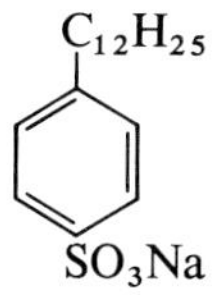

$C_{12}H_{25}$

SO_3Na

Fig. 2.3 Orthododecylbenzene sulphonate.

The hydrocarbon base is petroleum derived and does, in fact, contain a distribution of chain lengths with the predominant species being C_{12}. In addition, there can be a greater or lesser degree of chain branching. The sulphonation process utilized can vary from direct reaction with sulphuric acid to SO_2/SO_3 mixtures, but always results in some excess sulphuric acid. On neutralization, a proportion of sodium sulphate is produced which is preferably kept to a minimum for admixture formulations.

2.2.4 Alkyl sulphates

The literature describes the use of several materials of this type and these are shown in Table 2.2. These products are also compatible with many water-reducing agents to produce air-entraining water-reducing agents.

Table 2.2 Alkyl sulphates used as air-entraining agents

Material	Formula	Reference
Sodium dodecyl sulphate	$C_{12}H_{25}SO_4Na$	[7–11]
Sodium tetradecyl sulphate	$C_{14}H_{29}SO_4Na$	[12]
Sodium cetyl sulphate	$C_{16}H_{33}SO_4Na$	[8, 10, 11]
Sodium oleyl sulphate	$CH_3(CH_2)_7CH{=}CH{-}(CH_2)_8SO_4Na$	

2.2.5 Phenol ethoxylates

Phenol ethoxylates differ from the previous four categories in that they are non-ionic materials. Although not widely used, they are very effective at low addition levels and solutions of 2 to 4% by weight in water perform satisfactorily at low dosage level. The most common material is nonylphenol ethoxylate and limited studies [13] have indicated that the higher value of n in Fig. 2.4 is the most effective.

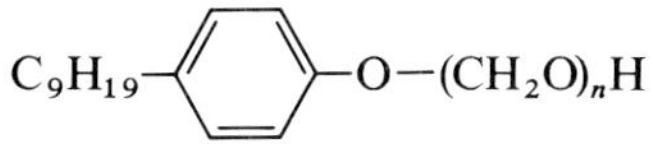

C_9H_{19}—⟨benzene ring⟩—$O{-}(CH_2O)_nH$

$n = 5$ to 15

Fig. 2.4 Nonylphenol ethoxylates.

Although the vast majority of commercially available air-entraining agents are simple solutions of materials within one of the above categories, it is possible to produce mixtures and this is occasionally done.

2.3 The effects of air-entraining agents on the water–cement system

Numerous studies have been made on the effect of additions of air-entraining agents to cement pastes which enable an insight to be gained into the mechanism by which these materials produce the stable microscopic air void system, and to some extent, their effect on the properties of concrete. However, unlike the water-reducing agents, the comparison of behaviour between that in cement pastes and that in total concrete systems, is not so useful because of the influence of the aggregate component in determining the air content and the resultant rheological characteristics. The data can be conveniently described under the following headings:

(a) Rheology.
(b) Air content and characteristics.
(c) Distribution between solid and aqueous phase.
(d) Effect on cement hydration reactions.
(e) Interpretation as a mechanism of action.

2.3.1 Rheology

The effect that air-entraining agents have on the rheology of fresh cement pastes can be considered from the point of view of changes due to the admixture itself, and those due to the presence of entrained air.

An interesting programme of work was carried out to isolate the individual effects [14] by preparing the cement paste/admixture mixes in two ways as given below.

(i) *Without air entrainment*

The apparatus shown in Fig. 2.5 was used which was filled completely with cement paste. Admixture additions were made by injecting with a hypodermic needle through the rubber cap. Pastes prepared in this way had air contents of less than 0.6% by volume, were free from any premature stiffening tendencies and were homogeneous.

(ii) *With varying degrees of air entrainment*

The same apparatus was used, but quantities of paste were removed to give

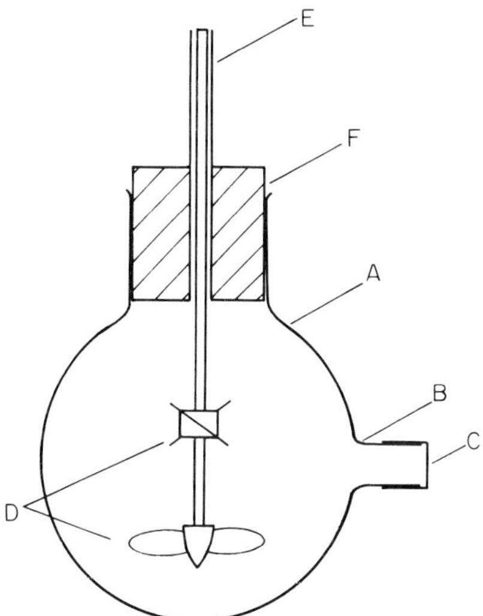

Fig. 2.5 Mixing apparatus for experiments on air-entrained and non-air-entrained pastes (Bruere).

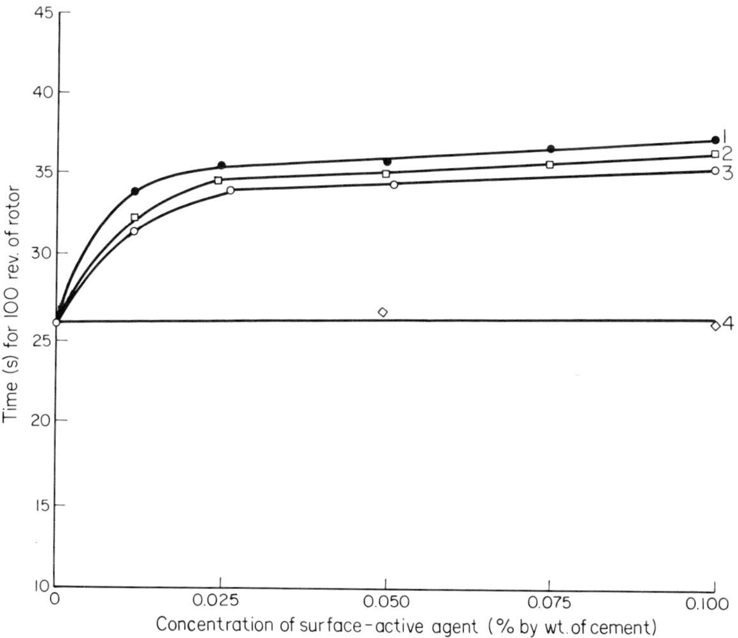

Fig. 2.6 Viscosities of cement pastes containing varying concentrations of surface active agents in the absence of entrained air (Bruere). 1 = sodium dodecyl sulphate; 2 = sodium abietate; 3 = petroleum suphonate; 4 = phenol ethoxylate.

an air space in the vessel. On rapid agitation the volume increased, dependent on the air content required.

Paste viscosities were measured, using a Stormer Viscometer, which is a type of concentric cylinder viscometer. Although it is possible to obtain results in absolute terms, for comparative purposes the times for 100 revolution of the rotor under a fixed applied torque were recorded.

In the absence of entrained air, the results shown in Fig. 2.6 were obtained where the similarity of behaviour of the three anionic materials, sodium dodecyl sulphate, sodium abietate and a petroleum sulphonate in increasing the paste viscosity is seen. The non-ionic material, phenol ethoxylate, has almost no effect on the paste viscosity when no air is entrained.

Table 2.3 The effect of air entrainment on paste viscosity for two different cements

Cement no.	Surface area (cm^2g^{-1}) (Blaine)	% Sodium abietate wt of cement	Time for 100 rev. of rotor (s)	% increase in time for 100 rev.
1	3600	0	26	38
1	3600	0.05	36	
2	4350	0	74	76
2	4350	0.05	130	

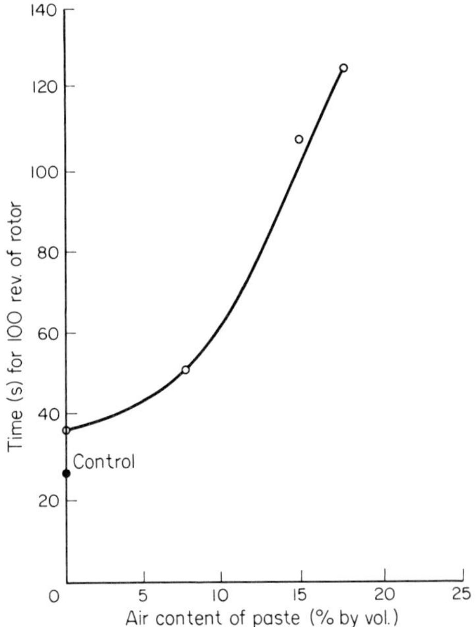

Fig. 2.7 The viscosity of a cement paste containing sodium abietate in the presence of entrained air (Bruere).

When sodium abietate was incorporated into pastes made from two cements having significantly differing suface areas, the results given in Table 2.3 were obtained, indicating that the sodium abietate produces a much greater increase in viscosity in paste made from the fine cement than in pastes made from the coarse cement.

When the viscosities of air-entrained pastes were measured by the same means, the results shown in Fig. 2.7 were obtained for sodium abietate at 0.05% by weight of cement. It can be seen that the magnitude of the effect due to the presence of the admixture itself is small in relation to the effect of the air it causes to be entrained.

2.3.2 Air content and characteristics

(i) *Dosage and admixture type*

The effect of varying the quantity of several air-entraining agents is shown in Fig. 2.8 [11]. The similarity of behaviour of sodium dodecyl sulphate and sodium abietate is again illustrated.

Table 2.4 summarizes similar data, but also includes the sodium dodecylbenzene sulphonate type and the resultant effect on the specific surface area and computed spacing factor of the bubbles [15].

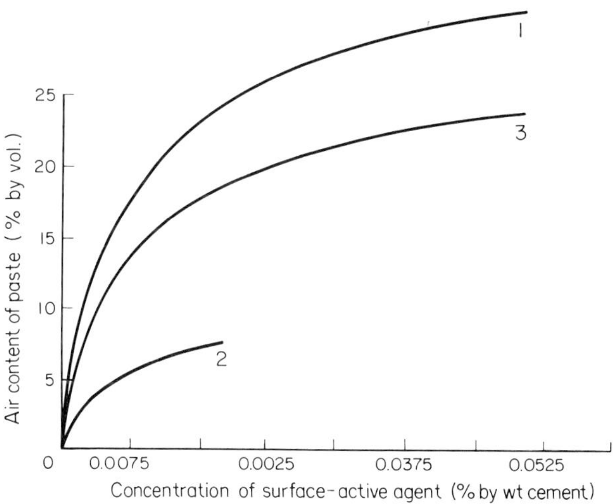

Fig. 2.8 Variations of air-entraining capacities of surface active agents in cement pastes with varying concentrations of agents (Bruere). 1 = sodium dodecyl sulphate; 2 = sodium tetradecyl sulphate; 3 = sodium abietate.

Table 2.4 The effect of various air-entraining agents at different concentrations on the specific surface area and computed spacing factor of air bubbles in cement paste

Surface active agent	Concentration (% by wt of cement)	Air content (% by volume)	Specific surface area of bubble ($mm^2\ mm^{-3}$)	Computed spacing factor (mm)
None		0.3	31	0.813
Sodium dodecyl sulphate	0.0025	6.5	68	0.112
	0.005	14.0	73	0.074
	0.010	17.5	78	0.066
	0.025	21.3	78	0.058
	0.050	27.2	78	0.048
Sodium dodecyl-benzene sulphonate	0.005	4.0	56	0.188
	0.010	5.6	61	0.135
	0.025	11.2	64	0.094
	0.050	13.1	61	0.091
Neutralized wood resins	0.005	4.6	53	0.170
	0.010	10.1	56	0.112
	0.025	17.4	53	0.094
	0.050	22.6	54	0.084

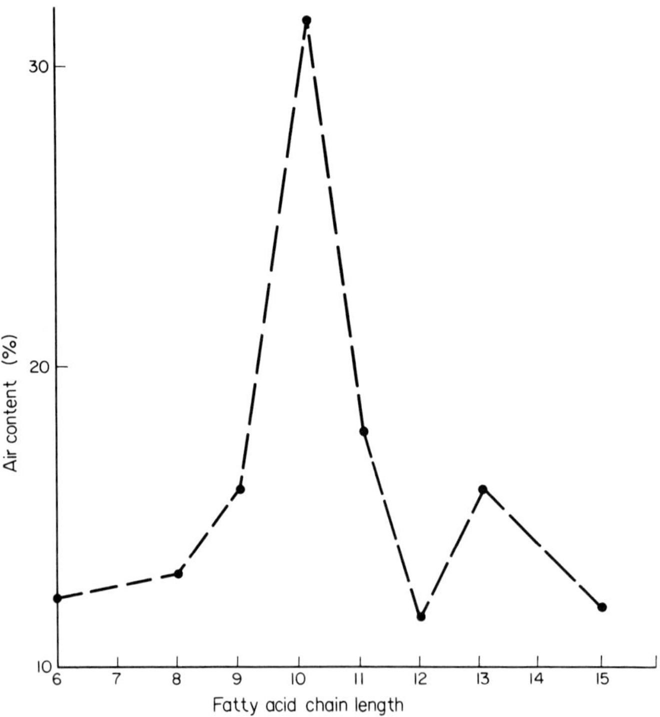

Fig. 2.9 The effect of increasing chain length of the sodium salts of linear fatty acids on air-entraining capacity (Rixom).

The specific surface area of bubbles entrained by each air-entraining agent is largely independent of its concentration and there are indications that sodium dodecyl sulphate, apart from being more effective as an air-entraining agent than either the sodium dodecylbenzene sulphonate and neutralized wood resins, also results in a high specific surface area of bubbles in closer proximity.

The effect of altering the fatty acid chain length of sodium salts of linear fatty acids on the air entrained in mortars is illustrated in Fig. 2.9 [6]. The superiority of the 9 to 11 carbon chain length fatty acids is clearly shown and, in practice, fatty acid fractions with a high C_{10} content are chosen; the C_9 and C_{11} fatty acids do not occur naturally in appreciable quantities.

(ii) *Water–cement ratio*

An increase in the water–cement ratio of cement pastes leads to greater air entrainment and a decrease in the specific surface area of bubbles. However, the spacing factor is relatively unchanged as shown in Table 2.5 [15].

(iii) *Cement type*

The characteristics of the cement used to prepare air-entrained cement pastes have a marked influence on the amount of air entrained and are

Table 2.5 The effect of water–cement ratio of cement pastes on the air content, specific surface area and computed spacing factor

Surface active agent	W/c ratio (by wt)	Air content (% by vol.)	Specific surface area of bubble ($mm^2 mm^{-3}$)	Computed spacing factor (mm)
0.025% sodium				
dodecyl	0.40	16.7	81	0.064
sulphate	0.45	21.4	69	0.066
	0.50	25.8	56	0.069

Table 2.6 The relationship between particle size of cement and the level of air entrainment

Particle size of cement (BSS mesh)	Air content of paste (% by vol.)
20–52	44.1
52–100	32.0
140–200	24.8
passing 200	21.0

Table 2.7 Variations in cement characteristics have a considerable effect on the level of air entrainment at constant water–cement ratio

Cement batch number	Surface area of cement (cm^2g^{-1}) (air permeability)	Total alkali content of cement	SO_4 content of cement	Air content of paste (% by vol.)
1	3300	1.09	1.99	14.7
2	3730	0.57	2.56	6.1
3	3480	0.12	1.85	11.0
4	3500	0.25	1.82	5.7

illustrated in Tables 2.6 and 2.7 for pastes prepared under standard conditions at a water–cement ratio of 0.45 [11].

The large differences, however, can largely be attributed to differences in viscosities of the different pastes at the same water–cement ratio. When the pastes were prepared to the same consistencies, similar levels of air entrainment were obtained.

(iv) *Temperature*

Over the range of 18 to 35°C, the effect of temperature is not great and is illustrated in Fig. 2.10 for a paste containing 0.0125% sodium dodecyl sulphate at a water–cement ratio of 0.45 [11]. It can be concluded that the degree of air entrainment obtained in cement pastes of a given consistency is largely a function of the type and dosage of air-entraining agent incorporated, whilst other variables have only a minor effect.

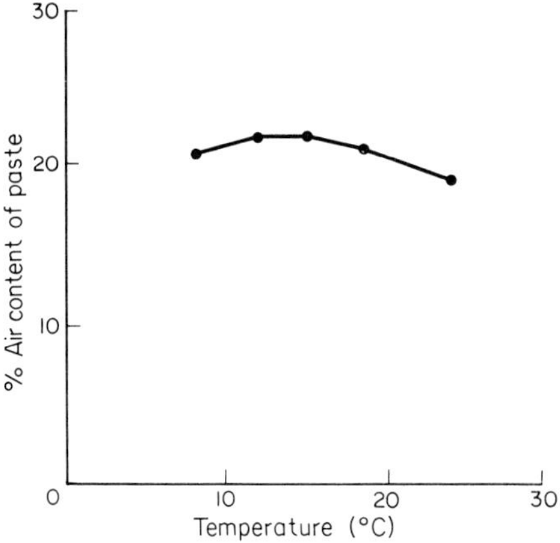

Fig. 2.10 The effect of temperature on the air-entraining capacity of sodium dodecyl sulphate in a cement paste (Bruere).

2.3.3 Distribution between solid and aqueous phases

A study of the foaming capacities and stabilities [11] of a variety of air-entraining agents in a solution of cement extracts, showed that commonly used anionic air-entraining agents, such as sodium dodecyl sulphate and sodium abietate (a) were visually precipitated from solution, (b) retained their ability to form stable foams after precipitation with only minor amounts of admixture left in solution, and (c) lost the major part of their ability to form stable foams after filtration. It was further shown from studies in cement pastes that (i) the admixture should be adsorbed on the solid particles of the paste with the non-polar ends of the molecule pointed towards the water phase. This imparts a hydrophobic character to the cement particle to which the air bubbles can adhere, and (ii) the residual concentration of admixture in the mixing water, although not necessarily high, must be sufficient to generate bubbles during mixing.

In cement paste, the residual concentration of calcium abietate and calcium resinate is approximately 0.5 g l^{-1} in the aqueous phase, which compares well with the aqueous solubility of calcium caprate, as shown in Fig. 2.11. It will be noted that the C_9 fatty acid has a solubility of about 1 gl^{-1} whilst the C_{11} is about 0.05 gl^{-1}. The solubility of calcium dodecyl sulphate is approximately 0.1 g l^{-1}, so it is indicated that the optimum value lies in the region of 0.1 to 0.5 g l^{-1} solubility of the calcium salt.

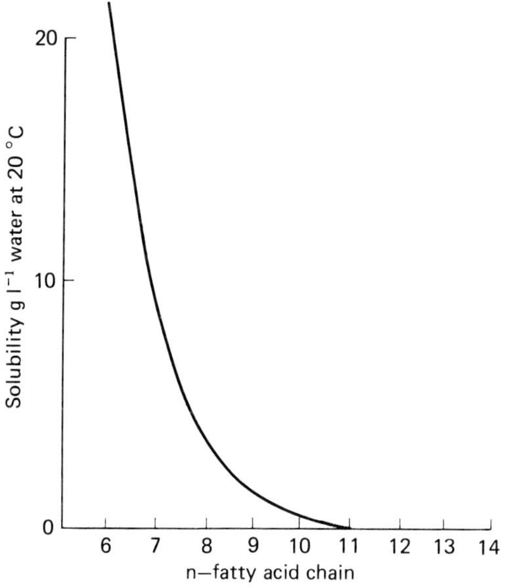

Fig. 2.11 The solubility of linear fatty acid calcium salts in water.

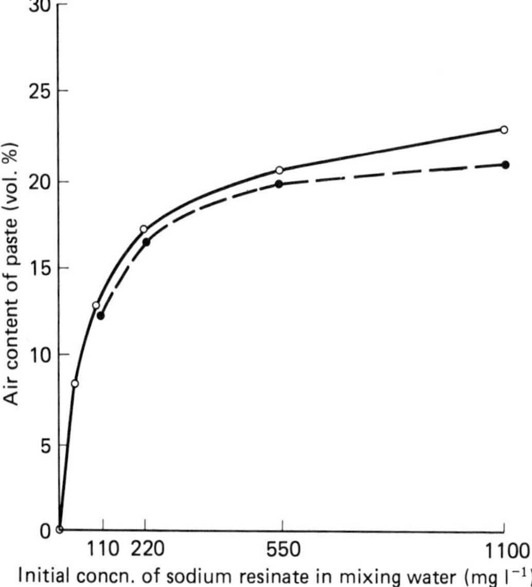

Fig. 2.12 Comparison of the air-entraining capacities of sodium and calcium abietates. ○= sodium abietate dissolved in mixing water and added to cement; ● = made with filtrates from mixtures of sodium abietate and cement paste extract (Bruere).

It is interesting to note that the abilities of the soluble sodium and insoluble calcium salts to entrain air in cement pastes are very similar [7] (Figs 2.12 and 2.13).

2.3.4 Effects on the hydration chemistry of cement

There is little published data on the effect of air-entraining agents on the chemistry and morphology of cement hydration. However, the limited studies [16] indicate that the normal hydration pattern under isothermal conditions for ordinary Portland cement shown in Fig. 2.14 is modified as follows:

(a) For sodium oleate based air-entraining agents, the C_3S peak is not affected, but the C_3A peak is accelerated and splits into two up to a 10 times normal dosage level. It is believed that the ettringite and monosulphate reactions are retarded due to an impermeable layer of a calcium oleate–aluminate hydrate salt.

(b) For the other anionic air-entraining agents, such as neutralized wood resins and sulphates or sulphonates, high dosages lead to a retardation

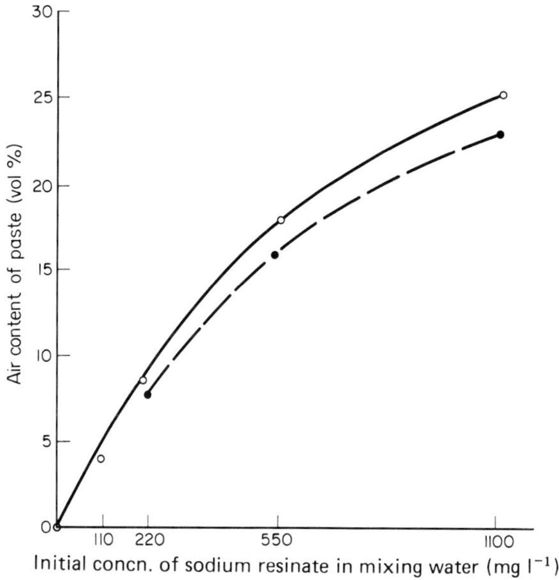

Fig. 2.13 Comparison of the air-entraining capacities of sodium and calcium resinates (wood resins). ○= sodium resinate dissolved in mixing water and added to cement; ●= made with filtrates from mixtures of sodium resinate and cement paste extract.

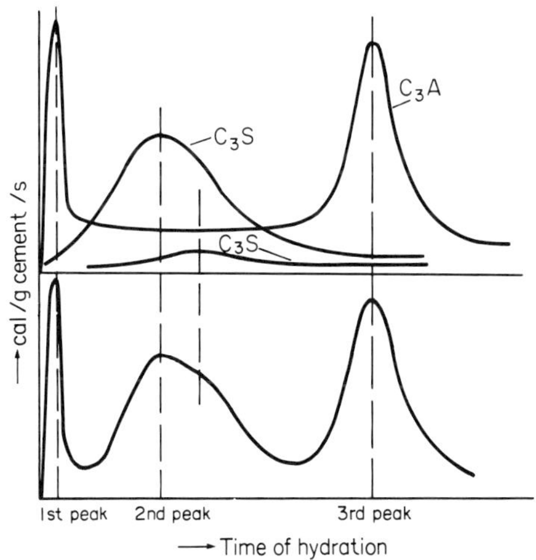

Fig. 2.14 Schematic diagram of the development of the heat of hydration of Portland cement under isothermal conditions (Bruere).

of the C_3S peak, whilst the C_3A peak is accelerated, and sometimes splits into two.

(c) Non-ionic materials, such as ethoxylates, do not appear to alter the heat output pattern of ordinary Portland cement.

Any changes in morphology of hydration products have not been published.

2.3.5 Interpretation as a mechanism of action

Air-entraining agents are predominantly anionic surfactants which, on addition to cement pastes, are adsorbed on to the cement particles with their polar groups orientated towards the particles. This 'sheath' is of limited solubility and only a minor, but finite, proportion remains in solution as the calcium salt.

The weak surfactant solution forms bubbles on agitation in the aqueous phase, which are stabilized as microscopic spheres from coalescing into large bubbles by the orientation of insoluble surfactant across the air/liquid interface and by adhering to the hydrophobic surface created on the cement particle by the adsorbed surfactant. This is shown diagrammatically in Fig. 2.15 [17].

The bubbles are less than 0.25 mm diameter and probably do not exist in the fresh paste with diameters less then 10 μm because the high pressure present in such small bubbles would cause the air to be dissolved.

Air-entrained pastes possess a higher viscosity than pastes with little or no air content, mainly because of the bridging effect of cement particles by air bubbles increasing the structure of the system.

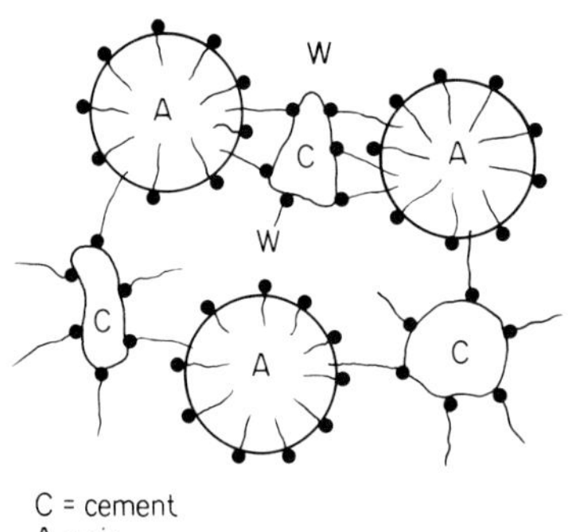

Fig. 2.15 The interactions between cement, air, water and molecules of air-entraining agent (Kreijger).

There is no evidence to suggest that the presence of air-entraining agents of the type normally available commercially alter, in any way, the eventual hydration products of the cement.

2.4 The effects of air-entraining agents on the properties of plastic concrete

In considering the effect that air-entraining admixtures have on the properties of concrete in the plastic state, it is necessary to consider the variables that influence the volume of air entrained, the stability of the air void system during any subsequent handling, placing and compaction processes, and the effect that the entrained air has on the workability, water content, mix stability and mix design procedures relevant to a concrete mix. It is clear that such differences could be considerable in view of the significant alteration that air-entraining agents make to the cement paste fraction of the cement; a normal addition level of air-entraining agent will incorporate an additional, say, 4% air by volume, but this air is made up of bubbles predominantly less than 100 μm diameter, only about 0.003 mm apart, and totalling some 250 000 cm^{-3} of paste.

2.4.1 Volume of air entrained

Most air-entraining agents are formulated to give a total of 3 to 6% air by volume in most normal concrete mixes at their recommended dosage levels. However, it is often found that the recommended starting dosage results in either too much, or too little entrained air, and adjustments have to be made. In some cases, the adjustment in dosage is considerable and can be due to a number of mix ingredient or environmental factors. Alternatively, the particular application may require less air entrainment (to improve mix stability) or more air entrainment (for density, or improved freeze–thaw resistance reasons) than the normal levels given below. In view of this, it is useful to have an understanding of the way the many variables can influence the volume of air entrained by this category of admixture, and these are discussed below.

(a) Dosage

In any mix, the greater the quantity of air-entraining agent added, the larger will be the volume of entrained air. Figs 2.16 and 2.17 [18, 19] indicate that there is a maximum air content possible by increasing the admixture dosage and Fig. 2.17, in particular, suggests that this maximum is dependent on the characteristics of the cement.

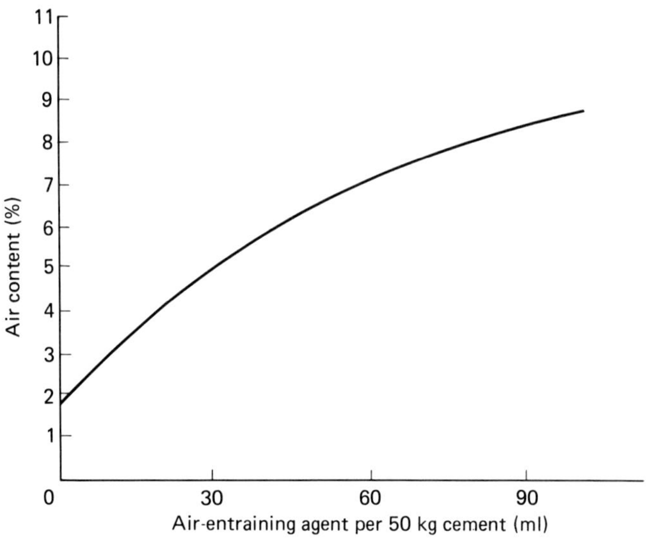

Fig. 2.16 Air content as a function of admixture addition level (Johnson).

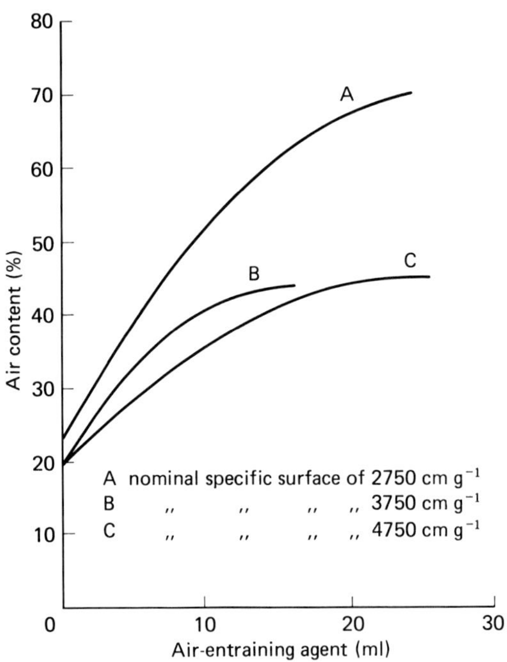

Fig. 2.17 Air content as a function of admixture addition level for cements of different fineness (Mayfield).

(b) Mixing techniques

The change in air content on prolonged mixing has been studied under laboratory conditions for two cements with results shown in Table 2.8 [20]. The data indicate that the maximum level of air entrainment is rapidly achieved in the case of the lower cement content mix and is progressively lost as mixing continues. In the case of the higher cement content mix, some 5 to 10 min. are required to reach the maximum air content which then diminishes on continued mixing. Field trials in a full-size mixer were carried out on the same admixtures using the 340 kg m^{-3} mix and the results are given in Table 2.9.

The two studies suggest that it is unlikely that more than about 1% of entrained air will be lost by mixing for up to 30 min.

The mix capacity was also varied using the full-size mixer with the same

Table 2.8 The effect of mixing time on the level of air entrainment in laboratory mixes

Mixing time (min.)	Air content (%)			
	No addition	Admixture		
		A	B	C
	256 kg m^{-3} cement			
2	1.8	5.3	6.0	3.8
5	1.1	5.2	5.5	3.4
10	1.0	4.2	4.9	2.1
15	1.0	3.3	4.2	1.9
30	1.0	2.5	3.4	1.1
60	1.0	2.0	1.5	1.0
	340 kg m^{-3} cement			
2	1.3	4.5	4.4	3.1
5	1.0	5.2	5.9	2.1
10	0.9	5.8	5.9	1.4
15	0.8	5.7	5.8	1.3
30	0.8	4.8	5.5	1.2
60	0.8	4.0	4.1	1.0

Table 2.9 The effect of mixing time on the level of air entrainment in a full-size mixer

Mixing time (min.)	No addition	Admixture		
		A	B	C
15	1.0	3.8	—	3.3
30	—	3.2	6.0	3.2
45	—	3.0	5.5	3.0
60	—	2.7	4.9	2.5
75	—	2.7	4.1	2.6
90	—	2.4	3.3	2.4

concrete mixes and admixtures. Results are given in Table 2.10 where it will be seen that the effect of batch size is only slight and, in the higher cement content mixes, shows a trend towards higher air content as the capacity of the mixes is approached.

(c) Cement characteristics

The characteristics and quantity of cement used in the production of air-entrained concrete can have a pronounced effect on the air content and/or the dosage of admixture required to obtain the necessary air content. This is illustrated by the evaluation [20] of twelve different cements in an identical mix using a standard dosage of admixture. Results are given in Table 2.11.

Large variations in air content due to a change in cement source are, therefore, quite possible and, although it is not possible to quantify the effect of all cement variables, the following data are relevant:

(i) *Fineness.* The fineness of cement is a major factor in determining the quantity of air-entraining admixture required to incorporate a given amount of entrained air. Table 2.12 summarizes the results for three cements differing only in their specific surface area [19]. Differences in cement fineness that are normally experienced could lead to a doubling or halving of admixture requirements, but it is worth noting that in the study from which the above data are derived, an addition level of 10.0 ml/batch would have produced concrete conforming to a 3 to 6% requirement for all three cements.

(ii) *Cement content.* The amount of entrained air decreases with increasing cement content [21] and typically, an increase in cement content of 90 kg m^{-3} will reduce the volume of entrained air in concrete by about 1% of the volume of concrete.

(iii) *Alkali content.* An investigation [22] of mortars containing cements differing only in their alkali content indicated that when the alkali (as Na_2O) in the water in contact with the cement reaches about 0.8% by weight of water, the amount of air-entraining agent required to achieve a given air content is minimized. This is shown in Fig. 2.18.

This effect also applies to alkali added as part of the admixture formulation and is particularly important in cements with low alkali analysis and is illustrated in Table 2.13 [23].

(d) Workability

If the addition level of the air-entraining agent is maintained at a constant level, a more workable mix will entrain more air than a less workable one.

Table 2.10 Slightly higher air contents are obtained as the volume of mix approaches that of the mix capacity, particularly for the higher cement content mix

% of mixes capacity	Air content (%)			
	No addition	Admixture		
		A	B	C
	$256\ kg\ m^{-3}$ cement			
20	1.5	5.1	5.1	2.7
40	1.6	6.1	5.1	4.2
60	1.5	6.1	6.0	3.8
80	1.5	5.8	6.0	3.5
100	1.5	6.0	6.2	3.9
	$340\ kg\ m^{-3}$ cement			
20	0.7	3.0	3.0	1.9
40	0.8	4.4	3.9	2.2
60	1.0	4.5	4.1	2.4
80	1.0	4.3	4.1	2.9
100	1.1	4.6	4.5	3.0

Table 2.11 Cement type and source can influence the volume of air obtained in both plain and air-entrained mixes

Cement code	Air entrained (%)			
	No addition	Admixture		
		1	2	3
A	3.1	5.2	5.3	3.6
B	1.2	5.3	4.4	3.2
C	1.2	5.4	5.5	3.0
D	1.1	5.1	5.4	2.0
E	1.4	3.9	3.5	2.8
F	2.3	6.2	5.4	4.5
G	2.4	7.9	6.3	4.3
H	1.0	5.4	4.5	3.6
I	1.3	6.2	5.2	3.0
J	1.8	7.8	8.1	4.6
K	1.0	6.9	5.9	3.6
L	1.7	7.2	5.8	4.3

Table 2.12 The amount of air-entraining agent required to obtain 4% air is increased for higher surface area cements

Cement	Specific surface area	Quantity of admixture required (ml/0.0368 m^3 batch) for 4% air
A	2750	6.5
B	3750	10.0
C	4750	14.0

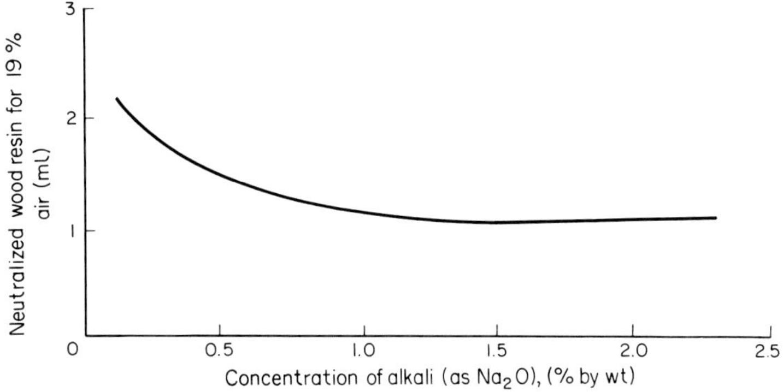

Fig. 2.18 The effect of alkali content of cements on the air-entraining capacities of neutralized wood resins in concrete (Greening).

Table 2.13 The alkalinity of the admixture influences the air content obtained, particularly with cements of low alkali analysis

Air-entraining admixture	pH	Na_2O(%)	Air content (%) of cement
1	9.4	0.35	5.5
2	10.6	0.38	7.4

However, for very workable concrete of slump greater than 180 mm, the air will be more rapidly lost before placing.

(e) Temperature

The temperature of the concrete has a considerable influence on the amount of air entrained by a standard addition level of admixture; the higher the temperature, the lower the air content. A typical set of results is shown in Fig. 2.19 [24].

(f) Aggregate type and content

(i) *Coarse aggregate.* There is no evidence to suggest that coarse aggregate shape or geological origin affect the amount of air entrainment obtained. The only exception is where crushed rock aggregates contain an appreciable quantity of dust which could influence the fine aggregate gradings considered below.

(ii) *Fine aggregate.* The amount of air that is entrained in concrete increases with an increase in the sand content and this is shown for a variety of sand gradings in two concrete mixes in Fig. 2.20 [25]. As a guide, an increase in

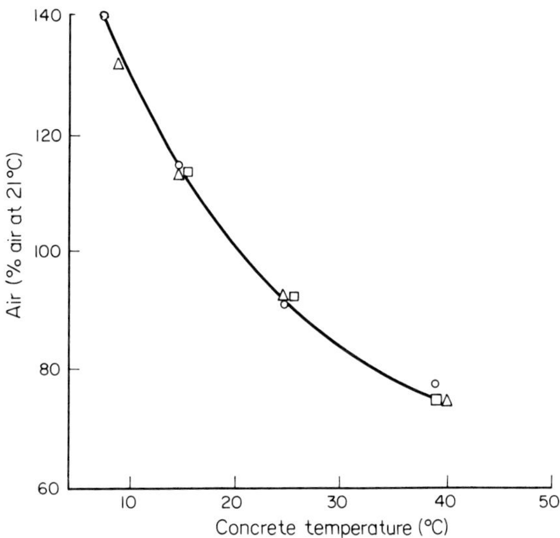

Fig. 2.19 The effect of temperature on the air entrainment of concrete at a standard addition level of air-entraining agent (Bloem).

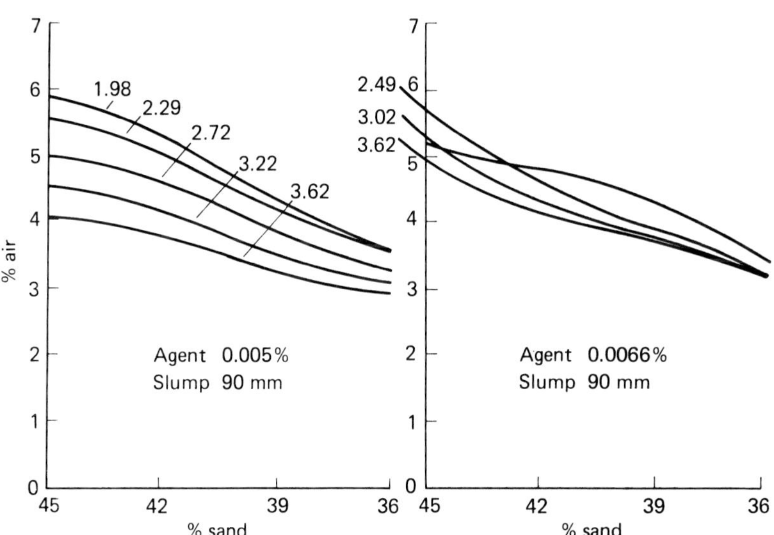

Fig. 2.20 The effect of sand content on the air entrainment of concrete at two addition levels of air-entraining agent, the fineness modulus (FM) of the sand also being varied (Craven).

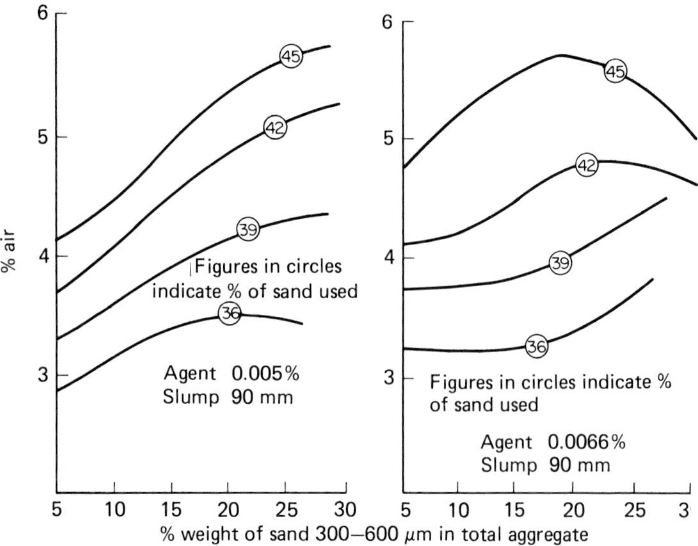

Fig. 2.21 The effect of fine aggregate of 3–600 micron diameter on the air entrainment of concrete (Craven).

the sand content of 5% will lead to an increase in air content of 1 to 1.5%.

A relationship exists between the size of bubbles that can be accommodated in a system containing fine particles, and the size of the particles. The size of the space subtended by particles of 300 to 600 μm diameter varies from about 30 to 100 μm. Since a large proportion of the bubbles in air-entrained concrete are less than 100 μm diameter, it is clear that this particle size is of considerable importance in determining the amount of air entrained. Therefore, even at constant sand content, an increase in the properties of particles of this size will lead to an increase in air content. This effect is shown for a number of sand contents in Fig. 2.21 [25].

(g) Fine fillers and pozzolans

When fine particle material of less than 20 μm diameter is included in the mix design, the amount of air-entraining agent must be increased to obtain the required air content [26]. The effect is largely independent of coarse aggregate type. Some results are shown in Fig. 2.22 where it is apparent that the effect is considerable in the case of fly ash and pumice [27]. Although these figures are relevant to a very high replacement of cementitious materials, even at normal replacement levels of 20 to 30% fly ash, the concrete will require three or four times the quantity of air-entraining agent in comparison to a concrete containing no fly ash to entrain the same volume of air.

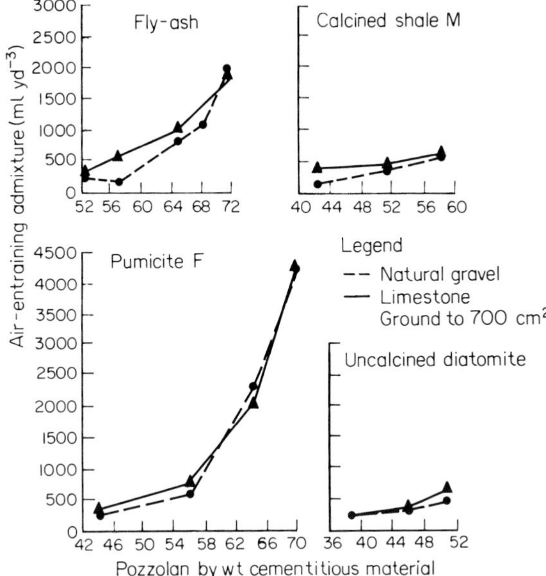

Fig. 2.22 The effect of pozzolan materials on the air-entraining agent dosage required to maintain the air content.

(h) The presence of water-reducing admixtures

If the concrete to be air entrained contains a water-reducing agent of the lignosulphonate or hydroxycarboxylic acid type, a reduction of 50 to 60% in the quantity of air-entraining agent can be made in comparison to a mix not containing the water-reducing agent [28, 29].

2.4.2 The stability of the entrained air

Following the removal of the concrete from the mixer, subsequent transport, handling and placing techniques can cause reductions in the air content and in the air void characteristics. However, it must be borne in mind that measurement of air content is usually by the pressure meter method which includes a compaction stage where the large voids that could be lost in handling operations are removed prior to measurement. In view of this, losses during handling and transport are usually less than 0.5% of the concrete volume as shown in Fig. 2.23 [30].

Losses of entrained air on vibrating placed concrete can be considerable, and typical results are shown in Fig. 2.24 [31]. The void size distribution is also altered and, as shown in Fig. 2.25 [31], the vibration mainly removes the larger diameter voids.

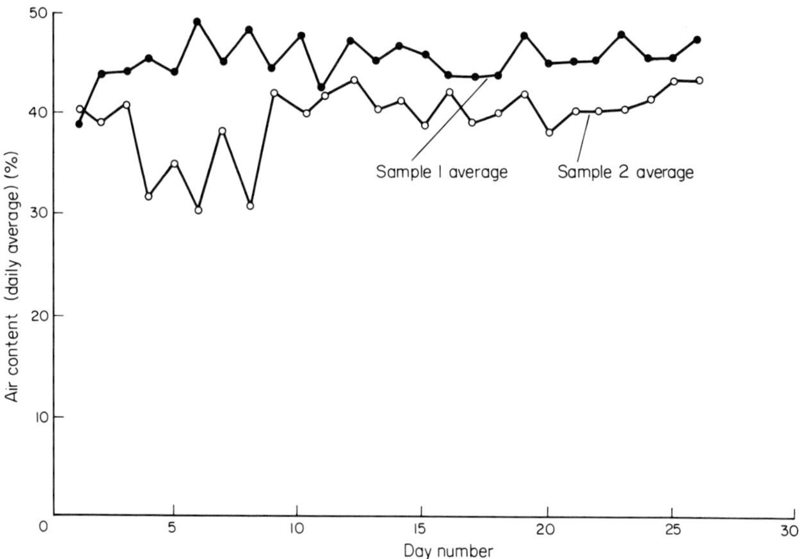

Fig. 2.23 Loss of air on handling of air-entrained concrete (McCurrich).

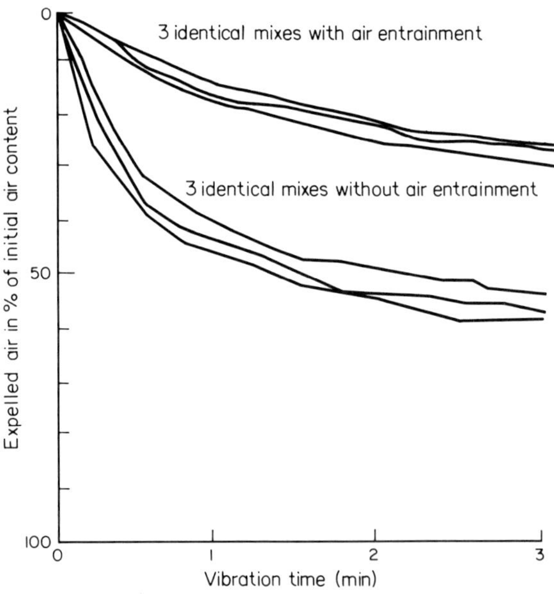

Fig. 2.24 Loss of air on vibration of air-entrained concrete (Johansen).

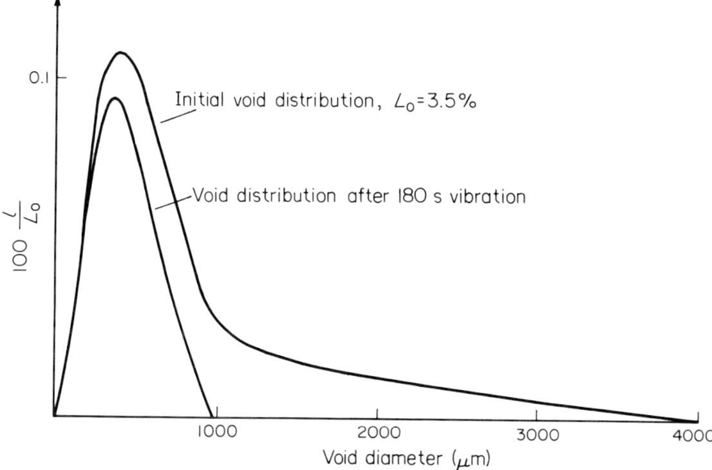

Fig. 2.25 Change in void size distribution on vibration of air-entrained concrete (Johansen).

In order to minimize air losses, air-entrained concrete should be compacted with the minimum of vibration, 5 to 15 s usually being sufficient.

2.4.3 Workability

The presence of entrained air results in an increase in the workability of concrete. This is in apparent contradiction with the paste thickening effect discussed earlier, and is due, in fact, to the increase in paste volume altering the paste aggregate volume ratio, and to the 'lubrication' of the coarse and fine agregate particles by the microscopic air bubbles. In terms of compacting factor and VeBe, some results are given in Fig. 2.26 [21] and the similarity of the slope of the lines suggests that for most mixes, an increase in air content of 5% of the concrete volume will result in an increase in the compacting factor of 0.06. A typical increase in slump would be from 12 mm to 50mm.

2.4.4 Water reduction

In order to maintain the original workability, a reduction in the water–cement ratio can be made of 5 to 15% depending on the air and cement contents. A typical relationship is shown in Fig. 2.27 [32].

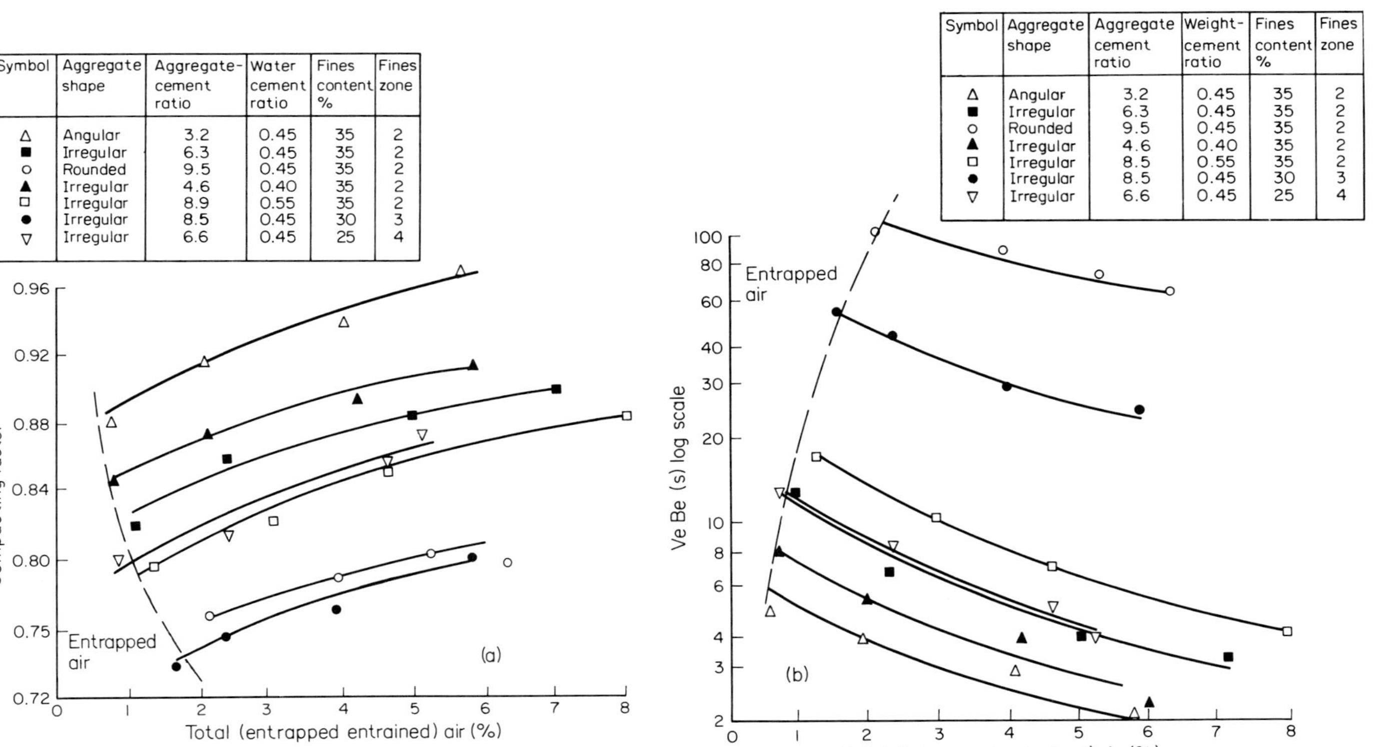

Symbol	Aggregate shape	Aggregate-cement ratio	Water cement ratio	Fines content %	Fines zone
△	Angular	3.2	0.45	35	2
■	Irregular	6.3	0.45	35	2
○	Rounded	9.5	0.45	35	2
▲	Irregular	4.6	0.40	35	2
□	Irregular	8.9	0.55	35	2
●	Irregular	8.5	0.45	30	3
▽	Irregular	6.6	0.45	25	4

Symbol	Aggregate shape	Aggregate cement ratio	Weight-cement ratio	Fines content %	Fines zone
△	Angular	3.2	0.45	35	2
■	Irregular	6.3	0.45	35	2
○	Rounded	9.5	0.45	35	2
▲	Irregular	4.6	0.40	35	2
□	Irregular	8.5	0.55	35	2
●	Irregular	8.5	0.45	30	3
▽	Irregular	6.6	0.45	25	4

Fig. 2.26 Effect of entrained air on (a) the compacting factor and (b) the VeBe times of various concretes (Cornelius).

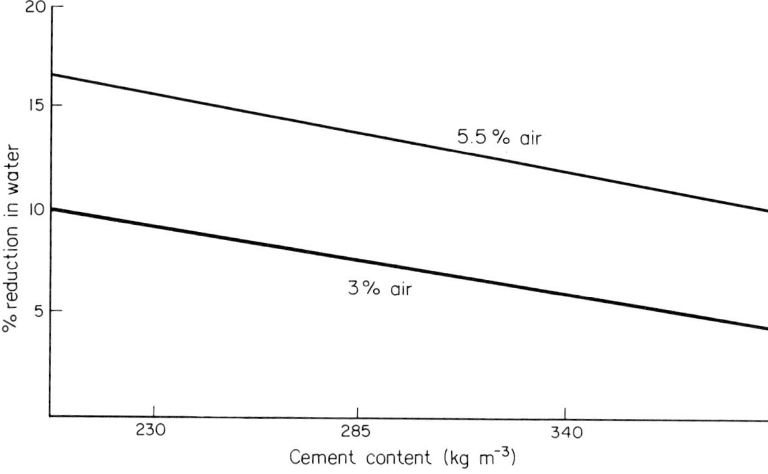

Fig. 2.27 Reduction in water content is dependent on the amount of air entrained and the cement content.

2.4.5 Mix stability

The presence of entrained air clearly makes the concrete more cohesive and allows a reduction in the quantity of fine aggregate, without increasing the tendency to segregate. The quantity of sand that can be removed from a concrete that is satisfactory in terms of cohesion prior to air entrainment, is approximately 20 kg m^{-3} for each 1% of additional air content.

Bleeding is always reduced by air entrainment, both under static conditions and during vibration, and is particularly useful for producing

Table 2.14 Bleeding of plain concrete, air-entrained concrete and air-entrained concrete containing a water-reducing agent

Water-reducing admixture	Air-entraining admixture	Slump (mm)	Air content (%)	Sand–aggregate ratio	ml cm^{-2} of surface	% total mix water
None . . .	none	87	1.6	0.45	0.48	14.2
None . . .	AEA-1	100	5.3	0.42	0.30	9.7
Class 1 . .	none	112	4.8	0.43	0.19	6.3
Product A .	AEA-1	112	5.5	0.42	0.17	6.1
Class 2 . .	none	112	4.7	0.43	0.25	8.6
Product B .	AEA-1	112	5.5	0.42	0.18	6.4
Class 2 . .	none	87	3.0	0.43	0.30	10.2
Product D .	AEA-1	100	5.4	0.42	0.23	7.8
Class 2 . .	none	100	3.8	0.43	0.28	9.5
Product E .	AEA-1	112	5.7	0.42	0.16	5.8

concrete from fine aggregate deficient in the lower particle size range where bleeding is often a problem. Some comparative data are given in Table 2.14 for plain mixes, mixes containing air-entraining agents, and mixes containing both air-entraining agents and water-reducing admixtures [33].

2.4.6 Mix design requirements

The addition of an air-entraining agent to concrete results in three changes which must be taken into consideration for mix design purposes: there is the increase in yield due to the volume of entrained air which will reduce the cement content per unit volume, the increase in cohesion allowing a reduction in sand content, and the increase in workability allowing a reduction in the water content. The following sequence of mix design procedure will accommodate these changes.

(i) Recognized mix design procedures (or existing standard mixes) are used to produce a plain concrete meeting the requirements of workability, strength and cohesive character using available materials.
(ii) The fine aggregate content of the mix is reduced by 20 kg sand/m^3 for each 1% of additional air required.
(iii) The air-entraining admixtures, at the manufactures' recommended addition level and pre-mixed in part of the gauging water, is added to the mix, and sufficient additional gauging water included to achieve the desired workability.

Table 2.15 Mix design procedure for air-entrained concrete

	Control mix	Modified air-entrained mix		
	(kg m^{-3})	1st adjustment (kg)		(kg m^{-3})
Composition				
20–5 mm gravel (rounded/angular)	1330		1330	1338
Sand (zone 2)	591	(−68)	523	526
Ordinary Portland cement	328		328	330
Water	181	(−15)	166	167
Air-entraining agent (ml)	nil		295	297
Properties of concrete				
W/c ratio	0.55		0.51	
Slump (mm)	38		45	
Air content (%)	1.8		5.5	
Compressive strength: 7 days	27.4		28.6	
(N mm^{-2}) 28 days	32.5		32.9	
Plastic density (kg m^{-3})	2430		2361	

(iv) Alterations in the admixture level may be necessary to achieve the required air content which will result in consequential changes in the gauging water requirements.
(v) Determine plastic density.
(vi) Correct batch figures for yield to obtain composition/m^3.
(vii) Prepare cubes for compressive strength determination.

This sequence is shown in Table 2.15, together with the resultant properties of the plastic and hardened concrete.

The combined effects of sand and water reductions bring the cement content to approximately that of the comparative plain concrete so that 28-day strengths are similar. However, this is not consistent across all cement contents; low cement content mixes tend to show increase in strength, whilst higher cement content will show a slight decrease. This is illustrated by the data in Table 2.16 [32].

Table 2.16 Effect of air-entrainment on strength

Total water–cement ratio	Plain concrete		Air-entrained concrete (4%)	
	Cement content* ($kg\,m^{-3}$)	28-day* strength ($N\,mm^{-3}$)	Cement content* ($kg\,m^{-3}$)	28-day strength* ($N\,mm^{-2}$)
0.44	383	35.0	341	29.4
0.49	341	31.5	298	27.3
0.53	312	28.0	284	25.9
0.58	284	25.2	256	21.7
0.62	270	22.4	241	19.6
0.67	256	19.6	227	17.5
0.71	241	17.5	213	15.4
0.75	227	14.0	199	14.0

*This is American data and although useful to show the comparative strengths for plain and air-entrained concrete, in practice, with higher cement contents and lower workability used in the UK, values should not be used as the basis for mix design.

2.5 The effects of air-entraining agents on the properties of hardened concrete

2.5.1 Structural design parameters

(a) Compressive strength

It is always assumed that the entrainment of air in concrete leads to a considerable reduction in compressive strength. However, it was shown earlier that because of mix design changes as a result of, or made possible by, the presence of the air bubbles, concretes containing 3 to 6% by volume of air and at constant cement content within the range 200 to 400 kg m^{-3}, will have only slight, if any, losses in strength. Indeed, at the lower cement content there could be an increase in strength (see Table 2.16).

It is useful, however, to be able to predict the 28-day strength of special concretes containing, perhaps, much higher air content for insulation or density reasons, or alternatively in mixes where no water or sand reductions are made. In these cases, the volume of additional air entrained should be considered in terms of an equivalent volume of water and added to the water already in the mix (after allowing for aggregate absorption). The new water and air/cement ratio can then be used to estimate the 28-day strength from standard curves of the type in Figs 1.51 and 1.52. Table 2.17 gives the change in water–cement ratio for calculation purposes for each 1% of air entrained.

Table 2.17 Factors to be added to the water–cement ratio to calculate the anticipated strength of air-entrained concrete

Cement content of plain mix (kg m^{-3})	Add to paste water–cement ratio of air-entrained mix to estimate strength for each 1% air entrained
100	0.100
200	0.050
300	0.033
400	0.025
500	0.020
600	0.017

(b) Flexural and tensile strength

There appears to be no published data on the effect of air entrainment of the tensile strength of concrete, but information is available on the flexural strength [34]. This is shown in Fig. 2.28 as a relationship between compressive strength and flexural strength for air-entrained and plain concrete mixes. This graph suggests that the same relationship exists for both types of concrete. However, other work has indicated that the flexural strength of an air-entrained concrete may be higher for a given compressive strength in comparison to a plain concrete [35]. The flexural strength is certainly not adversely affected.

(c) Stiffness

The relationship between the compressive strength and the modulus of elasticity for a number of similar concretes is shown in Fig. 2.29 [36] and Fig.

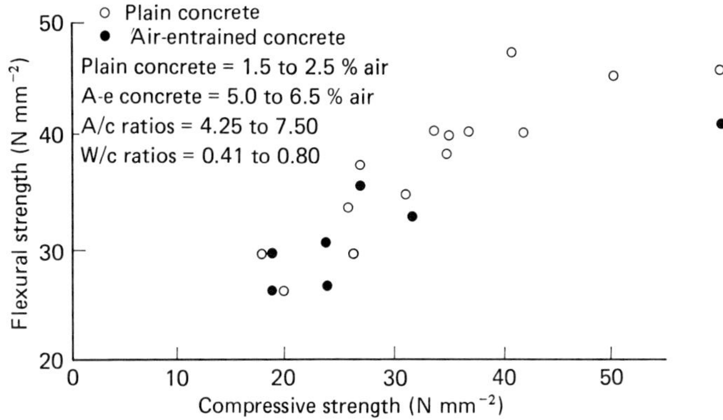

Fig. 2.28 The relationship between the flexural and compressive strengths of plain and air-entrained concretes (After Shacklock).

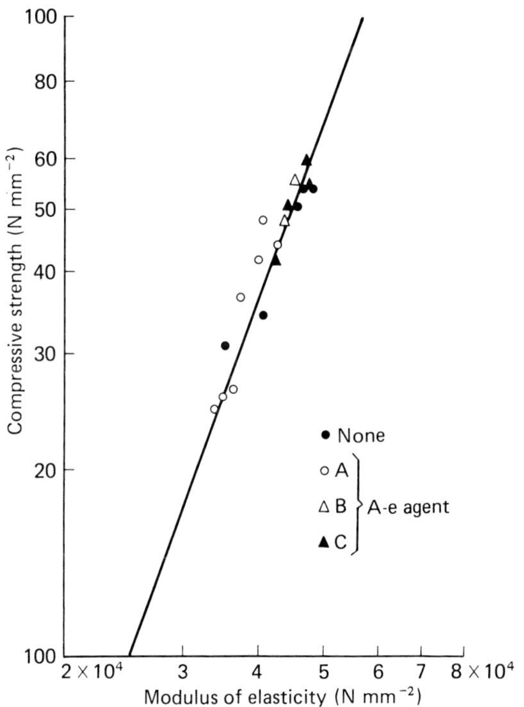

Fig. 2.29 The relationship between the compressive strength and modulus of elasticity of plain and air-entrained concretes (Warris).

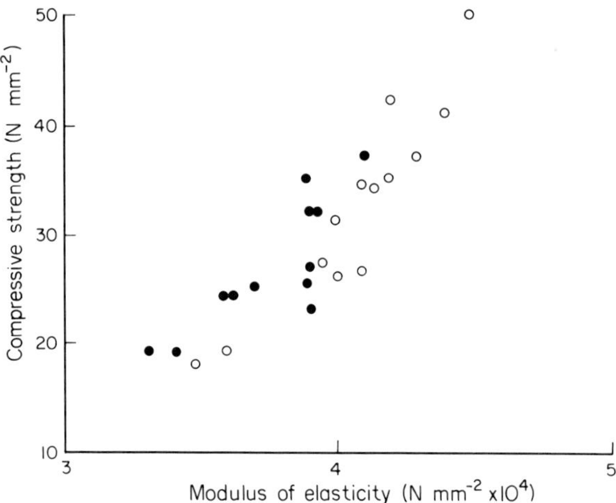

Fig. 2.30 The relationship between the compressive strength and modulus of elasticity of plain and air-entrained concretes (after Shacklock).

2.30 [34]. Again it is clear that the presence of entrained air does not alter the normal relationship.

The available data, therefore, suggest that by making the relevant mix design changes to compensate for the effect of entrained air on strength, the engineering properties of the concrete will be similar to a plain concrete of comparable strength.

2.5.2 Durability aspects

The important factors relevant to the ability of concrete to play its structural, aestheic and protective roles have been given earlier (Section 1.4). The effect that air-entraining agents have on these factors, with the exception of freeze–thaw resistance, has not been widely researched, but the available data are given below.

(a) Resistance to aggressive liquids

The permeability of concrete to aqueous liquids is reduced by most air-entraining agents [36] and this is illustrated in terms of the depth of penetration of water under a pressure of 8 N mm^{-2} in 48 h in Fig. 2.31. This is reflected in an improved resistance to attack by sulphate-bearing solutions which is indicated by the loss in ultrasonic wave velocity shown in Fig. 2.32 [37].

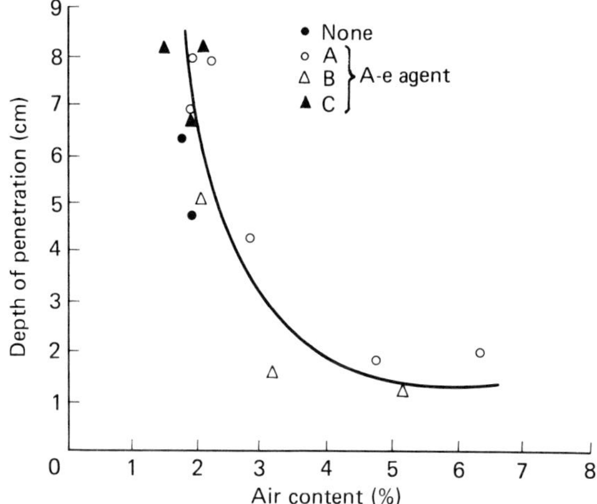

Fig. 2.31 The relationship between permeability and air content (Warris).

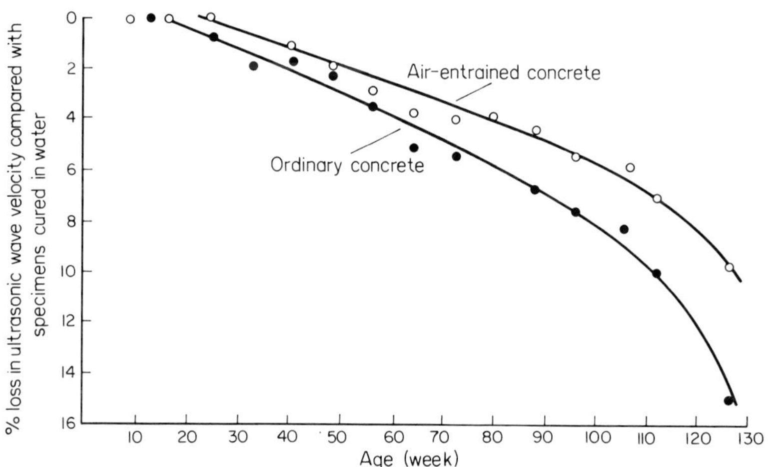

Fig. 2.32 Deterioration of plain and air-entrained concrete in sulphate solutions (Wright).

(b) Resistance to freeze–thaw conditions

The major application of air-entraining admixtures is to improve the resistance of concrete to freeze–thaw cycling particularly in the presence of de-icing salts. There is an abundance of data to illustrate the beneficial effect under these conditions, and it was shown earlier (see Fig. 2.1) that for any concrete of water–cement ratio greater than 0.45 that will be subjected

Table 2.18 Air entrainment improves the freeze–thaw resistance of concrete in the presence of de-icing salts

	Control	Air-entrained
Mix design		
10 mm gravel (kg)	8.0	8.4
Zone 2 sand (kg)	4.0	3.6
OPC (kg)	2.0	2.0
Properties		
Air content	1.0	4.5
Freeze–thaw resistance (number of cycles to cause 0.5 mg mm^{-2} scaling in 3% NaCl solution)	7.5	42.5

to freeze–thaw conditions, air entrainment is advisable. Table 2.18 summarizes the scaling, in the presence of de-icing salts, of specimens air entrained with a short chain fatty acid based air-entraining agent in comparison to a control concrete [38].

The degree of protection afforded is a function of the characteristics of the air void system entrained, and the significance of the variables involved requires an understanding of the mechanism of freeze–thaw resistance operating.

When water freezes it expands approximately 9% in volume and, if the expansion is restrained, a considerable force is exerted. The pressures exerted under total restraint are shown in Table 2.19 [30].

Table 2.19 Expansive pressures exerted by ice in a totally restrained system

Temperature (°C)	Pressure exerted in a totally restrained system (N mm^{-2})
0	0
− 10	119
− 20	190

In the hardened cement paste of a concrete mix, some of the original water put into the mix is held in a chemically bound form and is known as non-evaporable water. The remainder of the water, and any additional water taken up under saturated conditions resides in the gel and capillary pores and in any voids, and part of this 'evaporable' water is in a form that is able to freeze at low temperatures.

Table 2.20 gives an indication of the proportion of water than can form a solid phase at −20°C [39]. When this water freezes in the capillaries, it is

Table 2.20 The proportion of water that can form solid ice at − 20 °C as a function of paste water–cement ratio

Original w/c ratio of paste	% evaporable water able to freeze at − 20 °C
0.45	23
0.50	35
0.60	52

apparent that restraint against expansion is not total but, nevertheless, significant forces are set up within the capillary structure, as shown in Fig. 2.33 due to friction at the walls of the capillaries, and to end restraint due to the possibility of the system being 'full', or in the saturated state. These expansive forces need not be large to cause tensile failure (or spalling) in the concrete because of the relatively low tensile strength of concrete in the region of 2 to 6 N mm^{-2}.

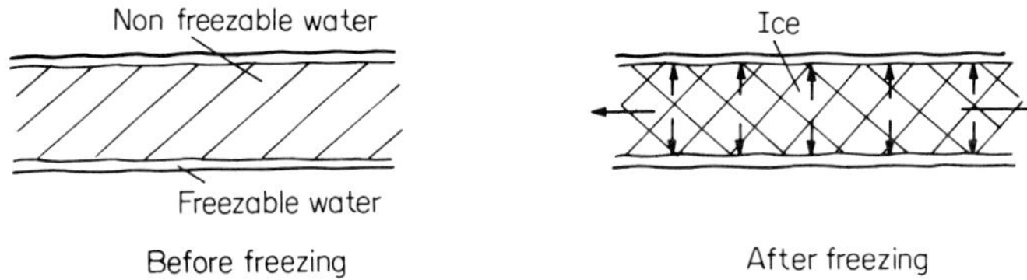

Fig. 2.33 Expansive forces are created within the capillary structure during freezing.

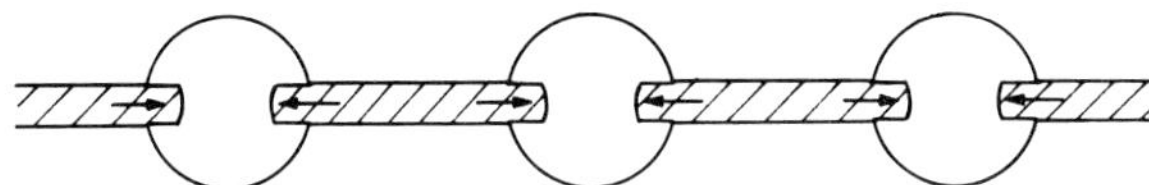

Fig. 2.34 The minute air bubbles act as reservoirs for ice expansion.

In air-entrained concrete, the minute air bubbles act as a reservoir for ice expansion as shown in Fig. 2.34. The effect is regulated by the following factors:

(i) *The amount of air space available in relation to the expansion of that part of the water likely to freeze.* In theory, the amount of void space necessary is about 4% of the paste volume but, in practice, because of the distribution of void sizes formed and because the total volume of the voids could not be filled by the ice which would already be in the solid form on entering the void, greater air contents are required. The recommendations shown in Table 2.21 are made.

Table 2.21 Recomended air contents for various maximum aggregate sizes

Maximum size of aggregate (mm)	Air content (% of concrete volume)
40	4 ± 1.5
20	5 ± 1.5
10	7 ± 1.5

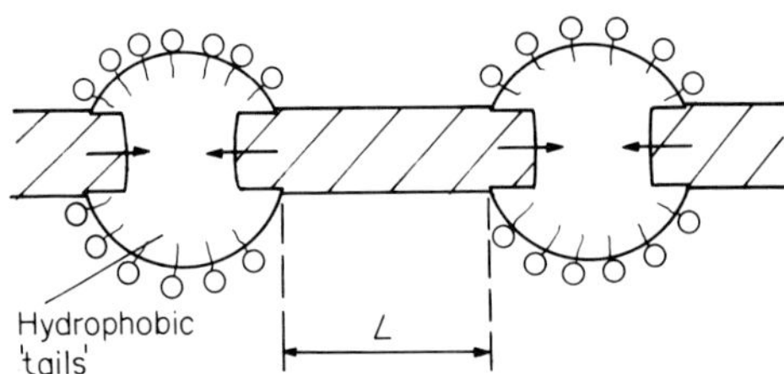

Fig. 2.35 The hydrophobic 'tails' of the air-entraining agent molecules prevent the filling of the air bubbles with water.

(ii) *The bubbles need to be empty and not full of water.* This is important to allow for expansion and this is ensured by orientation of the molecules of the air-entraining agent hydrophobic 'tails' into the voids as illustrated in Fig. 2.35, preventing the 'wetting' of the high contact angle void internal surface presented. When the ice expands into the void, the higher pressures easily overcome this effect. This explains, at least in part, why non-hydrophobic materials, like the phenol ethoxylates, although resulting in air entrainment, do not bestow significant improvements in freeze–thaw resistance to concrete in comparison to materials producing hydrophobic bubbles, such as the neutralized wood resins and fatty acid soaps (see Table 2.22 [40]).

Table 2.22 Freeze–thaw resistance of concrete air-entrained by different chemical types of admixture

Air-entraining admixtures	Air content (%)	Relative classification of freeze–thaw resistance*
None	2.0	5
Sodium oleate	5.6	86
Sodium lauryl sulphate	5.8	46
Neutralized wood resins	5.2	57
Phenol ethoxylate	5.2	7

*An average of several techniques measuring effect on modulus and compressive strength after freeze–thaw cycling.

(iii) *The length of the water column in the capillary up to a point where expansion is possible* (*L* in Fig. 2.35). This is to minimize the wall friction to allow the freezing water column to move and is a function of the spacing factor of the air void system (the average distance between adjacent air bubbles). It has been shown that the separation should be at least 0.4 mm to be at all effective, and preferably nearer to one-fifth to one-tenth of this value. Again, this is an important factor in explaining differences in the performance of air-entraining agents of different types, as shown in Table 2.23.

Table 2.23 The effect of spacing factor on freeze–thaw durability

Air-entraining admixture	% Air by volume	Average spacing factor (mm)	Durability to freeze–thaw cycling*
None	1.0	1.01	17
Non-ionic air-entraining agent	5.0	0.35	24
Neutralized wood resin	5.1	0.15	70

*An average of several techniques measuring effect on modulus and compressive strength after freeze–thaw cycling.

(iv) *Rate of freezing*. This is important in allowing time for the freezing water column to exude from the capillary into the void and so the more rapid the fall in temperature, the more severe the effect. This is the major reason for the severity of damage in the presence of de-icing salts, sodium and calcium chloride; the melting of ice at the surface causes latent heat to be extracted from the top layer of the concrete, resulting in very rapid freezing in the capillaries near the surface. The thermal gradient in the concrete leads to the worst tensile stress situations.

It should be pointed out that deterioration under freeze–thaw conditions can also be caused by a mechanism other than the direct freezing of the non-evaporable water. The capillaries contain dissolved salts, such as hydroxides, sulphates and carbonates. As part of the water is frozen, the concentration of salts in the remaining water increases and water will flow by osmotic pressure from the gel pores to the capillary pores, setting up an additional disruptive pressure.

(c) The protection of steel reinforcement

The formation of the passive layer at the concrete/reinforcement interface referred to earlier (Section 1.4) is due to the alkaline nature of the concrete.

Table 2.24 Depth of carbonation of air-entrained and plain concrete

	Concrete type					
	OPC/sand and gravel 350 kg m^{-3}			OPC/lightweight aggregate 350 kg m^{-3}		
Admixture	None	Neutralized wood resins	Petroleum sulphonate	None	Neutralized wood resins	Petroleum sulphonate
Atmospheric exposure*	1.00	0.69	0.65	3.65	1.55	3.19
Indoor exposure*	1.00	0.79	0.48	2.29	1.75	1.38

*The figures are expressed as relative to the control mix OPC/sand and gravel.

The alkalinity is due to calcium, sodium and potassium hydroxides which, over a period of time, react with atmospheric carbon dioxide to form carbonates. This reduction in alkalinity in reflected in a diminished protective capacity towards the steel reinforcement.

Some carbonation tests have been reported on plain and air-entrained concretes using two aggregate types, where the depth of carbonation on indoor and atmospheric exposure for 5 years have been measured [41]. The results are given in Table 2.24. In all cases, the air-entrained mixes have

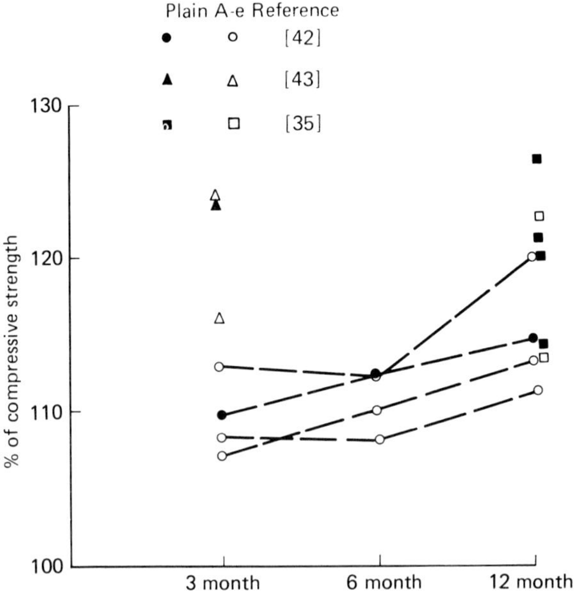

Fig. 2.36 The increase in compressive strength of plain and air-entrained concrete up to a period of one year.

Table 2.25 Drying shrinkage of air-entrained concrete is no greater than that of plain concrete

Air-entraining admixture	Concrete proportions				Air %	Slump (cm)	Drying shrinkage $\times 10^{-6}$		
	Cement ($kg\,m^{-3}$)	Water ($l\,m^{-3}$)	W/c ratio	Fine agg. (%)			7 days	28 days	56 days
None	300	150	0.50	40	1.7	6	220	546	688
A	300	135	0.45	36	4.5	6	225	560	689
B	300	135	0.45	36	4.4	6.5	253	586	698
C	300	135	0.45	36	4.7	6.5	215	525	661

shown less carbonation than similar controls, suggesting that air-entrained concrete should provide a better reinforcement protection in the long term.

(d) Maintenance of compressive strength

In a previous section it was shown that by suitable changes in mix design, concrete could be produced containing extrained air with similar 28-day compressive strengths to plain concrete. There are no published data, however, for the strength development of air-entrained concrete over considerable periods of time. Neither, of course, is there any evidence to the contrary and, indeed, the relatively small number of recorded compressive strengths of up to 1 year indicate no fall off in strength in comparison to control concretes as shown in Fig. 2.36 [35, 42, 43].

Experiments on the effect of different curing conditions on the compressive and flexural strengths of plain and air-entrained concrete [34] showed that air-entrained concrete has less tendency to lose moisture under drying conditions, which means that if concrete curing conditions are not ideal, air-entrained concrete should develop strength more normally than plain concrete [44].

(e) Volume deformations

(i) *Shrinkage.* Air-entraining agents do not increase the drying shrinkage of concrete, and mixes designed to have similar strength and workability characteristics can be considered to have similar drying shrinkage, whether air entrained or not. Typical data are shown in Table 2.25. This has been confirmed by other workers [36, 45, 43].

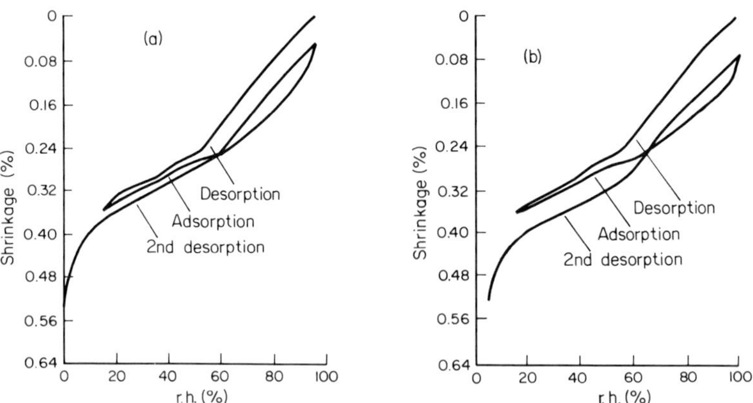

Fig. 2.37 Length changes (shrinkage) of plain and air-entrained cement pastes (Feldman).

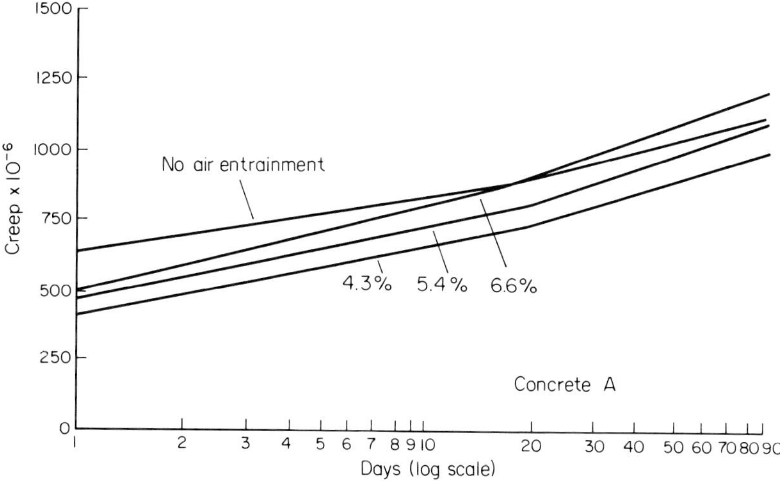

Fig. 2.38 The creep of plain and air-entrained concretes at 21 °C and a stress–strength ratio of 45% (Nasser).

A programme of work on cement pastes [46] showed that subsequent rewetting and drying in the presence of neutralized wood resins showed no increase in shrinkage in comparison to a plain paste. The results are given in Fig. 2.37.

(ii) *Creep*. The creep of air-entrained concrete at a constant stress–strength ratio is said to be not greater than that of comparative plain concretes [47]. However, at higher air contents (greater than 6%) there may be a very marginal increase in creep after some time under load, although at earlier ages this trend does not appear to be observed [48] Fig. 2.38 illustrates this effect.

Air-entraining admixtures, therefore, produce concrete which is more durable to conditions of freezing and thawing, particularly in the presence of de-icing salts, more resistance to sulphate attack, provides better protection to embedded reinforcement and is more tolerant of poor curing conditions. There appears to be no great difference in the way air-entrained concrete behaves in terms of compressive strength development and volume deformations.

References

1 Jackson, F.H. (1944). *ACI Journal,* **14,** 509.
2 Della-Libera, G. (1967). *Proceedings of the International Symposium on Admixtures for Mortar and Concrete,* Brussels, 142.

3 Neville, A.M. (1963). *Properties of Concrete,* Pitman, London.
4 Anon (1965). *Admixtures for Concrete.* CVR Report No. 31, 36–7.
5 Hewlett, P.C. (1978). Private communication.
6 Rixom, M.R. (1975). *Proceedings of the Workshop on the use of Chemical Admixtures in Concrete,* University of New South Wales, 149–76.
7 Bruere, G.M. (1971). *Journal of Applied Chemistry,* **21,** 61–4.
8 Taylor, W.H. (1974). *Precast Concrete,* **5,** 83–4, 89–90, 96.
9 Bruere, G.M. (1967). *Proceedings of the International Symposium on Admixtures for Mortar and Concrete,* Brussels, 8–13.
10 Vivian, H.E. (1960). *Proceedings of the Fourth International Symposium on the Chemistry of Cement,* Washington, 917–8.
11 Bruere, G.M. (1955). *ACI Journal,* **51,** 905–19.
12 Bruere, G.M. (1960). *Australian Journal of Applied Science,* **11,** 289–94.
13 Anon (1965). *Admixtures for Concrete,* CVR Report No. 31, 45–7.
14 Bruere, G.M. (1958). *Australian Journal of Applied Science,* **9,** 349–59.
15 Bruere, G.M. (1967). *Proceedings of the International Symposium on Admixtures for Mortar and Concrete,* Brussels, 14–22.
16 Kreijger, P.C. (1967). *Proceedings of the International Symposium on Admixtures for Mortar and Concrete,* Brussels, 27–32.
17 Kreijger, P.C. (1967). *Proceedings of the International Symposium on Admixtures for Mortar and Concrete,* Brussels, 33–37.
18 Johnson, D.L. (1968). *ACI Journal,* **65,** 402–41.
19 Mayfield, B. (1969). *Civil Engineering and Public Works Review,* **1,** 37–41.
20 Scripture, E.W. (1949). *ACI Journal,* **45,** 653–62.
21 Cornelius, D.F. (1970). *Report LR363,* Road Research Laboratory.
22 & 23] Greening, W.R. (1967). *JPCA Research and Development Laboratories,* **45,** 22–36.
24 Bloem, D.C. (1946). *ACI Journal,* **42,** 629–39.
25 Craven, M.A. (1948). *ACI Journal,* **44,** 205–15.
26 Anon (1974). ACI Publications SP.46, 99–108.
27 Anon (1962). *Army Engineer Waterways Experiment Station,* Mississippi, No. AD 756 299.
28 Larson, T.D. *et al.* (1963). *ACI Journal,* **60,** 1739–53.
29 Mielenz, R.C. (1968). *Proceedings of the Fifth International Symposium on the Chemistry of Cement,* Tokyo, 10–18.
30 McCurrich, L.H. (1976). Private communication.
31 Vivian, H.E. (1960). *Proceedings of the Fourth International Symposium on the Chemistry of Cement,* Washington, 915.
32 Anon (1965). *Grace Technical Bulletin No. 8.*
33 Vollick, C.A. (1959). *ASTM Special Publication* No. 266, 194–5.
34 Shacklock, B.W. (1959). *Civil Engineering and Public Works Review,* **54,** 77–84.
35 Kreijger, P. (1954). *Lecture notes,* University of Delft, Holland.
36 Warris, B. (1967). *Proceedings of the International Symposium on Admixtures for Mortar and Concrete,* Brussels, 11–15.
37 Wright, P.J.F. (1953). *Proceedings of the Institute of Civil Engineering,* **2,** 337–58.
38 Rixom, M.R. (1975). *Chemistry and Industry,* **7,** 162–5.
39 Powers, T. C. (1945) A working hypothesis for further studies of frost resistance of concrete, *proc. ACI,* **41,** 245.
40 Kieijger, P.C. (1967). *Proceedings of the International Symposium on Admixtures for Mortar and Concrete,* Brussels, 237–44.

41 Nishi, T. (1967). *Proceedings of the International Symposium on Admixtures for Mortar and Concrete,* Brussels, 112–7.
42 Kobayashi, M. (1967). *Proceedings of the International Symposium on Admixtures for Mortar and Concrete,* Brussels, 80–2.
43 Pais-Cuddou, M. (1967). *Proceedings of the International Symposium on Admixtures for Mortar and Concrete,* Brussels, 191–205.
44 Sutherland, A. (1974). *Air Entrained Concrete.* Publication 45 022, Cement and Concrete Association.
45 Keene, P.W. (1960). *Technical Report TRA 331,* Cement and Concrete Association.
46 Feldman, R.F. (1975). *Cement and Concrete Research,* **5,** 25–35.
47 Anon (1970). *Shrinkage and Creep in Concrete.* ACI Bibliography No. 10, 1966–70.
48 Nasser, K.W. (1973). *Behaviour of Concrete under Temperature Extremes.* ACI Publications SP **39,** 139–48.

3
Concrete waterproofers

3.1 Background and definitions

Concrete waterproofers are integral admixtures which alter the concrete surface so that it becomes water repellent, or less 'wettable'. This is illustrated in Fig. 3.1 where a close-up of a water drop on a treated and plain concrete surface is shown.

Fig. 3.1 Waterproofed concrete exhibits a high contact angle to water.

Fig. 3.2 Waterproofed concrete bricks exhibit almost no capillary rise. 2A = Waterproofed. 2 = No waterproofer.

This water repellency conferred on the concrete is only effective in preventing water from entering the surface when the applied pressure is small, e.g. rainfall in windy conditions, or capillary rise. The latter effect is shown in Fig. 3.2. In view of this, these materials are used normally for improving the quality of bricks, blocks and cladding panels where the additional benefits of reduced efflorescence, the maintenance of clean surfaces and the more even drying out of adjacent bricks and panels are also obtained.

In water-retaining structures or basement concrete subject to high hydrostatic pressure, materials of this type are generally not beneficial. However, some waterproofing admixtures do contain water-reducing admixtures and will result in a reduction in permeability under an applied hydrostatic head. In addition, the reduced capillary size and quantity will increase the hydrostatic pressure required to enter the concrete surface (see later).

3.2 The chemistry of concrete waterproofers

The chemical materials used to produce waterproofers are able to form a thin hydrophobic layer within the pores and voids and on the surfaces of the concrete in one of three ways: (a) reaction with cement hydration products; (b) coalescence from globular particle form (emulsion); (c) incorporation in a very finely divided form. Table 3.1 summarizes the chemical types used in each category.

Table 3.1 Chemical types of concrete waterproofers

Method of hydrophobic layer deposition	Materials used in formulation	Reference
(i) Reaction with cement hydration products	Stearic acid $C_{17}H_{35}COOH$	
	Oleic acid $C_{17}H_{33}COOH$	[1]
	Vegetable and animal fats	[2]
	Butyl stearate	[4, 9]
	Caprylic ($C_7H_{15}COOH$) and Capric ($C_9H_{19}COOH$) acids	[3]
(ii) Coalescence from emulsion	Wax emulsion	[2, 5]
(iii) Finely divided form of material	Calcium stearate	[6, 9]
	Aluminium stearate	[8]
	Hydrocarbon resin	[7]
	Bitumen	[8]

3.2.1 Materials which react with cement hydration products

Waterproofers based on liquid fatty acids, such as oleic, caprylic and capric, are used as major components in fatty acid mixtures. A typical example is shown in Table 3.2 [3]. The mixtures are added directly to the concrete mix without pre-dilution, or addition to the gauging water.

Stearic acid is widely used in this application and can be added directly to the mix in powder form, pre-mixed with an inert filler, such as talc or silica, which aids dispersion throughout the mix or, for the same reason, as an emulsion in water.

Butyl stearate is also added as an emulsion and because of its slower reaction with the hydration products, it is claimed that better dispersion throughout the mix is obtained, so that less material is required, than in the case of stearic acid.

Certain vegetable and animal fats have been used as waterproofers and

Table 3.2 Typical composition of liquid fatty acid waterproofers

Fatty acid	(%)
$C_5H_{11}COOH$	1.1
$C_7H_{15}COOH$	73.3
$C_9H_{19}COOH$	21.5
$C_{11}H_{23}COOH$	4.1

Table 3.3 Typical composition of fat-based waterproofers

	Paste by weight	
Component	Emulsion	Paste
Finely divided silica	11.2	30.0
$Ca(OH)_2$	2.0	5.0
$CaCl_2$	2.2	10.0
$CaCO_3$	1.0	1.0
Fat	2.0	20.0
Water	25.0	20.0
Mineral spirits	—	10.0

again emulsions or pastes are preferred and typical formulations are given in Table 3.3 [10]. The fat can be white grease, tallow or soya bean oil and although they all produce hydrophobic concrete, different effects on compressive strength are obtained.

3.2.2 Materials which coalesce on contact with cement hydration products.

Very finely divided wax exulsions are effective concrete waterproofing agents and are formulated so that the emulsion breaks down after contact with the alkaline concrete environment and forms a hydrophobic layer. Waxes of melting point 57 to 60°C are used with an emulsifying agent based on sorbitan monostearate or ethoxylated sorbitan monostearate [2]. The properties of a commercial product are given in Table 3.4 [5].

Table 3.4 Typical characteristics of wax emulsion type of waterproofer

Appearance	Milky white emulsion
Specific gravity	0.98
pH	6.5–7.0
Viscosity (cP)	6–8
Size of wax particles (μm)	0.5–1.0
Solids content (%)	30

3.2.3 Finely divided hydrophobic materials

This type of waterproofing admixture is widely used in the precast concrete industry, in particular the calcium and aluminium stearates. The calcium stearates can be produced by grinding stearic acid with lime or cement to produce a material containing 10 to 30% calcium stearate.

Inert materials, such as hydrocarbon resins and coal tar pitches in fine powder form, are claimed [7, 8,] to maintain their ability to produce hydrophobic surfaces, even after autoclaving. They would typically be ground to pass the 200 μm sieve.

3.3 The effects of waterproofers on the water–cement system

There is not a great deal of published data on the effect that waterproofing admixtures have on cement pastes in the way that water-reducing and air-entraining agents have been studied, but some limited points can be made.

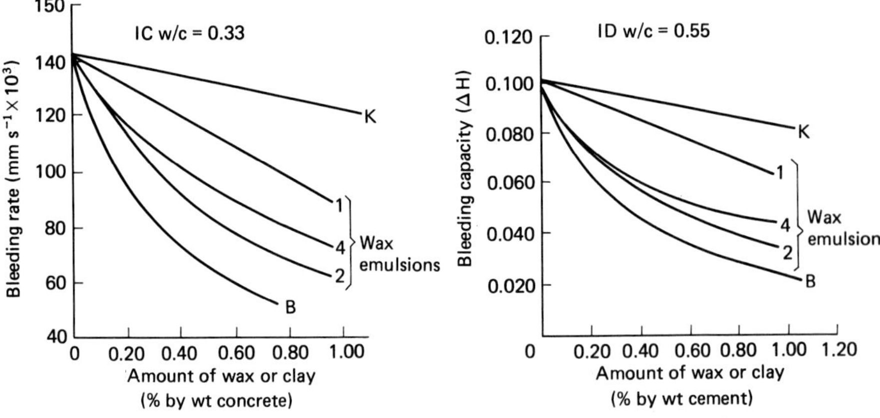

Fig. 3.3 Wax emulsions reduce the bleeding rate of concrete (Bruere).

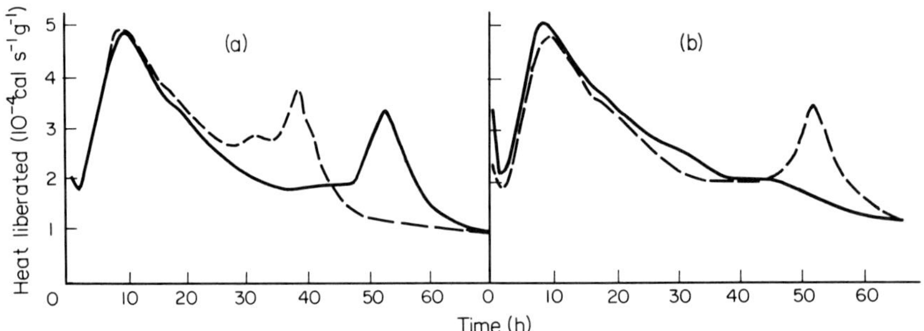

Fig. 3.4 The influence of sodium oleate on cement hydration (Edwards).

3.3.1 Bleeding of cement pastes

Wax emulsions have been shown [11] to cause a considerable reduction in the bleeding rates and capacities of cement pastes, and results are given in Fig. 3.3 in comparison to bentonite (B) and kaolin (K) additions. Although not as effective as bentonite, some of the wax emulsions are clearly very beneficial in this role.

3.3.2 Hydration of cement pastes

Addition of waterproofers based on caprylic, capric or stearic acid, stearates or wax emulsions do not have any effect on the setting characteristics of hydration products of Portland cement. However, the unsaturated fatty acid salts, such as oleates, although not affecting the tricalcium silicate hydration, have a marked effect on the ettringite and monosulphate reaction [12] and this is illustrated in the isothermal calorimetry results in Fig. 3.4. It is possible that a calcium oleoaluminate hydrate complex is formed involving the double bond of the oleic acid.

3.3.3 Effects on the capillary system of hardened paste

Hardened Portland cement contains a distribution of pore and capillary sizes, depending on the initial water–cement ratio and the maturity of the paste.

A typical distribution of pore radii in the hardened cement paste of concrete was shown in Fig. 1.56 which indicated that the majority of pores lie in the region of 0.05 and 1.0 μm diameter and it is through these pores that water passes by applied pressure or capillary rise as shown in Fig. 3.5(a).

It is believed that in the presence of waterproofing admixtures, the surfaces of the concrete, and the internal surfaces of the pores become coated with either a layer of molecules in the case of stearic acid and other fatty acids (Fig. 3.5b) or a layer of coalesced or separate particles of material

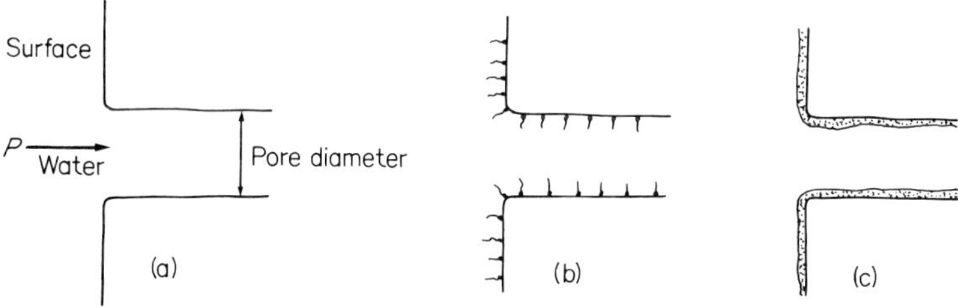

Fig. 3.5 The molecules or particles of the waterproofing admixture line the capillaries with a hydrophobic sheath.

in the case of waxes and bitumens, etc. (Fig. 3.5c). The end result in both cases is to produce hydrophobic surfaces exhibiting high contact angles to water as shown in Fig. 3.6.

In Fig. 3.5(a), the pressure P required to force water into the surface is given by the expression:

$$P = \frac{-2\gamma \cos\theta}{r}$$

where γ is the surface tension of water $= 72$ dyn cm^{-1}, θ is approximately 120° for surfaces coated with waxes or fatty acids, r is the radius of capillary $= 0.5\,\mu$m for the largest.

$$\therefore \quad P = \frac{-2 \times 72 \times \cos 120^\circ}{0.5 \times 10^{-4}} \text{ dyn cm}^{-2}$$

$$= \frac{2 \times 72 \times 0.5}{0.5 \times 10^{-4}} \text{ dyn cm}^{-2}$$

$$= \frac{2 \times 72 \times 0.5}{0.5 \times 10^{-4} \times 10^3} \text{ cm head of water}$$

$$= 1400 \text{ cm head of water approximately.}$$

The effect of the high contact angle surface is two-fold:

(a) The pressure required to enter the surface is positive, therefore, capillary rise should be nil. In fact, because of incomplete coating, there may be some slight rise in moisture, but this will be considerably reduced in comparison to an untreated concrete.

(b) The approximate requirement of a 14 m head of water to penetrate the surface through the largest capillaries can be related to the pressure exerted by the impact of the average raindrop in various wind conditions [13] shown in Table 3.5.

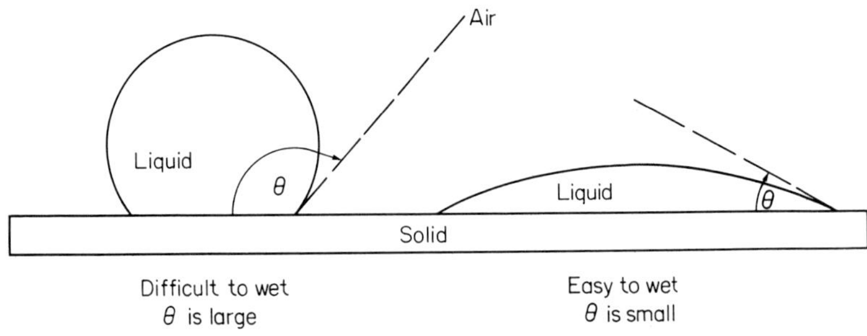

Fig. 3.6 High and low contact angles.

Table 3.5 Impact pressures of average raindrops at various wind speeds

Wind speed (km h^{-1})	Impact pressure of average raindrop of 0.05 g mass (cm head of water)
10	140
20	280
40	560
60	840
80	1120
100	1400
120	1680

Waterproofed concrete, therefore, should not show significant uptake of surface water in conditions of rain and wind up to about 100 km h^{-1}. In fact, on prolonged exposure, some wetting does occur, because of defects in the hydrophobic coating, and the presence of larger voids in the concrete, up to perhaps,1 mm wide; these are due to incomplete compaction, or to the nature of the concrete in the case of blocks.

3.4 The effects of waterproofers on the properties of plastic concrete

Waterproofing admixtures are formulated to affect the properties of the hardened concrete, and not those of concrete in its plastic state. In the case of materials based solely on calcium and aluminium stearates, stearic acid in solid or emulsion form, bitumens and hydrocarbon resins, there will be no effect on the properties of the plastic concrete with regard to air content, workability, mix design parameters, etc. When water-reducing admixtures or accelerators are included in the formulation, the effect on the concrete will be a function of the particular type of material used (see relevant section). The wax emulsions do appear to have an effect on the properties of the plastic concrete because of the 'lubrication' effect of the very small particles of wax, and the types of emulsifying agent used result in air entrainment in the region of 4 to 5% by volume.

3.5 The effects of waterproofers on the properties of hardened concrete

3.5.1 Structural design parameters

(a) Compressive and tensile strength

Admixtures of this category do not significantly affect the strength of concrete at 28 days unless a water-reducing admixture has been included.

In the formulation the resultant lowering of the water–cement ratio

Table 3.6 Effect of a wax emulsion type of waterproofer on the compressive strength of concrete of varying cement content

Mix type	Cement content ($kg\,m^{-3}$)	Compressive strength ($N\,mm^{-3}$) at 28 days
Plain	450	64.9
3% wax emulsion on cement weight	450	62.3
Plain	400	53.6
3% wax emulsion on cement weight	400	54.0
Plain	360	44.0
3% wax emulsion on cement weight	360	42.4
Plain	315	33.5
3% wax emulsion on cement weight	315	33.5
Plain	270	21.1
3% wax emulsion on cement weight	270	22.4
Plain	225	Too harsh to prepare cubes
3% wax emulsion on cement weight	225	15.3

will give a higher strength according to Abram's law. Some results for a wax emulsion type [14] at constant cement content and giving slight air entrainment (approximately 4% by volume) are given in Table 3.6, whilst Table 3.7 gives data for a dry brick mix containing a stearic acid based material.

(b) Modulus of elasticity

There are no recorded data to indicate that materials of this type would alter the stiffness of the concrete into which they are incorporated. However, the fact that these materials are associated with the matrix/air interface, and not the cement hydrates themselves, would suggest that the physical properties of the bonding constituents of the hardened cement would remain unchanged.

Table 3.7 Effect of a stearic acid based waterproofer on the compressive strength of concrete

Mix	% waterproofer on cement	Average compressive strength ($N\,mm^{-2}$) 7 days	28 days
1	0	14.2	17.6
2	1	14.3	18.6
3	2	15.2	18.2

Mix proportions: cement: sand: crushed limestone = 1 : 4.0 : 6.5, W/c ratio = 0.45 : 1

Table 3.8 The relationship between the initial surface absorption test and durability of concrete containing various proportions of a stearic acid based waterproofer

% waterproofer wt/wt cement	Initial surface absorption at 28 days (BS 1881) 10 min	30 min	1 h	Description after 10 years as coping on a roof site
0	0.48	0.28	0.20	Broken up due to frost, blackened by algae
0.10	0.42	0.25	0.19	Darkened by algal growth
0.25	0.45	0.20	0.17	Slightly darkened by algal growth
0.50	0.43	0.19	0.13	Slightly dirty
1.00	0.33	0.14	0.12	Fairly clean
2.00	0.07	0.04	0.02	Pristine condition

3.5.2 Durability aspects

It was explained earlier that materials of this type are added to concrete products, predominantly, but not necessarily, precast items, to reduce the ingress of rain and ground water for aesthetic and damp-proofing reasons, rather than to prolong the structural capabilities of the construction. The

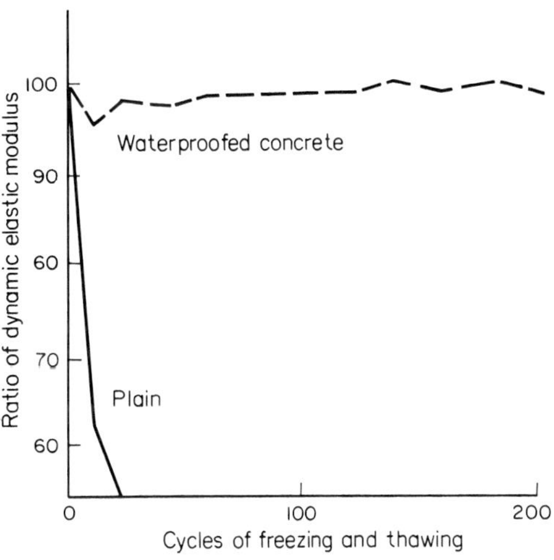

Mix	$kg\,m^{-3}$ cement content	W/c ratio	Air (%)
Plain	285	0.70	0.9
Waterproofed	285	0.61	4.2

Fig. 3.7 The freeze–thaw resistance of plain concrete and concrete waterproofed with a wax emsulsion type of admixture.

improvements in aesthetic qualities are not short lived, and results [15] for a stearate based composition over a 10 year period are given in Table 3.8.

The results given in Table 3.8 illustrate the following points:

(a) It is necessary to add sufficient waterproofer so that the absorption of the surface is reduced to a negligible level to obtain the best results.
(b) The presence of the waterproofing admixture at higher levels completely inhibits disfiguring algal growth.
(c) The presence of even small amounts of waterproofer improve the freeze–thaw durability of the concrete [16]. This is further indicated by freeze–thaw testing of concrete specimens containing the wax emulsion type of waterproofer shown in Fig. 3.7. In this case, however, some air entrainment and a reduction in the water–cement ratio was obtained which would contribute to the effect [5].

Volume changes such as shrinkage and creep are not significantly affected on single drying, as shown in Table 3.9 [5], but because subsequent moisture uptake is considerably reduced, the shrinkage under drying and wetting cycling (site conditions) will show a reduction.

Table 3.9 Drying shrinkage of concrete containing varying proportions of a wax emulsion based waterproofer

Specimen no.	% wax emulsion	Drying shrinkage of concrete					
		3 day	7 day	14 day	28 day	56 day	91 day
1	0	1.05	2.13	2.87	4.68	6.98	8.45
2	1	0.96	2.10	2.65	4.66	6.82	8.14
3	2	0.98	2.06	2.47	4.79	6.42	8.00
4	3	0.75	2.09	2.88	4.61	6.61	7.92

In view of the way in which these materials function, no improvement in resistance to attack by aggressive gases, e.g. industrial atmospheres, is obtained, although the reduced absorption of aqueous media will improve resistance to attack by aggressive, but neutral media, such as sulphates and other salts. Indeed, the presence of materials such as stearates at high level can completely inhibit the corrosion of reinforcing steel against a fairly high level of chlorides [17] in the concrete.

The waterproofing admixtures will, therefore, improve the aesthetic qualities of concrete in terms of maintenance of a clean appearance over a prolonged period of time without adverse effects on other properties, and in the areas of freeze–thaw resistance, shrinkage under wet–dry cycling and reinforcement protection, may contribute beneficially.

References

1 Nurie, R.W. (1953). *Cement and Lime Manufacture,* **26,** 47–51.
2 Australian Patent (1964). 271 527.
3 British Patent (1976). 1434 924.
4 Hewlett, P.C. Private communication.
5 Anon (1970). *Onada Technical Bulletin,* Onada Cement Company, Tokyo.
6 Lea, F. M. (1956). *Chemistry of Cement and Concrete,* 602–4, Chemical Publication Co. Inc., N.Y.
7 Tar Residuals Ltd. Private communication.
8 Dennis, R.H. (1970). *Chemistry and Industry,* 377–80.
9 Anon (1965). *Admixtures for Concrete,* CUR Report No. 31, 39–40.
10 Dory, D.M. (1969). *Cement and Lime Manufacture,* **42,** 107–14.
11 Bruere, G.M. (1974). *Cement and Concrete Research,* **4,** 557–66.
12 Nasser, K.W. (1973). *Behaviour of Concrete under Temperature Extremes,* ACI Publication SP 39, 139–48.
13 Wakenham, H. *et al.* (1945). *American Dyestuffs Report,* **18,** 178–82.
14 Brown, L.C. (1970). *Paper presented at the 50th Anniversary Convention of NCMA,* Texas.
15 Levitt, M. (1971). *British Journal of Non-Destructive Testing,* **12,** 106–12.
16 Shacklock, B.W. (1971). *Supplement to the Consulting Engineer,* **27,** 9–13.
17 Gouda, U.K. (1970). *British Corrosion Journal,* **5,** 204–8.

4
Accelerators

4.1 Background and definitions

Concrete accelerators increase the rate of hardening of cement and concrete mixes. The major material used to obtain this effect, calcium chloride, has been used since 1885 [1] and finds application mainly in cold weather, when it allows the strength gain to approach that of concrete cured under normal curing temperatures. In this way, not only can shutter and mould stripping, and lifting and handling of precast items, proceed normally, but also the concrete is less liable to damage by early age freezing. In this latter respect it should be pointed out that calcium chloride is not an anti-freeze in the sense of significantly lowering the freezing point of water in the mix, so that although the time required for protection is reduced, the standard methods of protection should be followed for cold weather concreting.

Although the use of accelerators is far greater in winter, they do find application under more normal conditions to speed up the setting and hardening process for earlier finishing or mould turn round.

There has been controversy over the use of calcium chloride in concrete containing embedded metal in view of the possibility of corrosion, particularly where the concrete is of a porous nature. More recently, the UK has followed the recommendations of many other countries and made provision in the relevant codes of practice to prevent its use where steel reinforcement is present. This has renewed interest in 'chloride-free' accelerators as replacements for calcium chloride in reinforced concrete. However, calcium chloride remains a most effective material for use in unreinforced concrete for economic production under winter conditions and its effects on concrete, whether beneficial or undesirable, are well researched and quantified. In some areas the newer non-chloride materials, although shown to reduce the likelihood of reinforcement corrosion, have not been widely studied and their other effects on concrete are largely unknown at the time of writing.

The fact remains that calcium chloride has been widely used as an accelerator for plain unreinforced concrete and this area of application, which in 1975 accounted for over 60% of calcium chloride usage, will continue in the future.

4.2 The chemistry of accelerators

This category of admixture is based on two major raw materials, calcium chloride and calcium formate [2] with minor amounts of other materials occasionally being included in the formulations, such as calcium nitrate, calcium thiosulphate [3] and triethanolamine (TEA). TEA is not normally used alone but because it is sometimes used in other categories of admixture to compensate for retarding influences it will be included in this section.

4.2.1 Calcium chloride

Calcium chloride ($CaCl_2$) is produced as a by-product in the Solvay process for sodium carbonate manufacture. The overall process involved is:

$$\underset{\text{limestone}}{CaCO_3} + \underset{\text{brine solution}}{2\,NaCl} \rightarrow Na_2CO_3 + CaCl_2.$$

It is obtained either as a liquor or as flake material of approximately 20% moisture content, and for use as an accelerating admixture is normally supplied as a 33 to 35% solution.

4.2.2 Calcium formate

Calcium formate ($Ca(HCOO)_2$) is produced as a by-product in the manufacture of a polyhydric alcohol, pentaerythritol:

$$\underset{\text{acetaldehyde}}{CH_3CHO} + \underset{\text{formaldehyde}}{3\,HCHO} \xrightarrow{Ca(OH)_2} HO{-}CH_2{-}\underset{CH_2OH}{\overset{CH_2OH}{\underset{|}{\overset{|}{C}}}}{-}CHO$$

$$\downarrow HCHO \mid Ca(OH)_2$$

$$\underset{\text{calcium formate}}{Ca(HCOO)_2} + HO{-}CH_2{-}\underset{CH_2OH}{\overset{CH_2OH}{\underset{|}{\overset{|}{C}}}}{-}CH_2OH$$

pentaerythritol

It is obtained as a fine powder and is supplied normally in this form as an accelerating admixture because of its limited solubility in water (about 15% at normal room temperature).

4.2.3 Triethanolamine

Triethanolamine ($N(C_2H_4OH)_3$) is an oily, water-soluble liquid with a 'fishy' odour and is produced by the reaction between ammonia and ethylene oxide:

$$NH_3 + 3\,\overset{O}{\overbrace{CH_2-CH_2}} \longrightarrow N(CH_2CH_2OH)_3$$

It is normally used as a component in other admixture formulations and rarely, if ever, as a sole ingredient.

4.3 The effects of accelerators on the water–cement system

4.3.1 Rheological effects

Most admixtures of this type do not significantly alter the rheology of cement pastes at early ages. The quicker stiffening of accelerated pastes will, of course, result in higher viscosities at a later age. More complex formulations occasionally include water-reducing admixtures to reduce the water–cement ratio, and their effect will be a function of the water-reducing admixture type and content (see Section 1.3.1).

4.3.2 Chemical effects

Calcium salts

The reactions between calcium chloride and the constituents and reaction products of Portland cement have been widely researched and are of importance in practice since the risk of corrosion of reinforcement depends, at least in part, on the amount of chloride which is left in a free state in solution in the concrete [4].

The following points are relevant to the reactions occuring in pastes of Portland cement containing normal proportions of C_3A, C_3S, C_2S, C_4AF and gypsum in the presence of calcium chloride.

(a) There does not appear to be any chemical reaction between calcium chloride and the di- and tri-calcium silicates [5] although their rate of reaction is increased.

(b) Calcium chloride does not react significantly with cement pastes for a period of 2 to 6 h [1,5] after mixing, although rapid setting can occur in this period.

(c) The free calcium chloride concentration, after the initial period described in (b), progressively drops to almost nil.

(d) The formation of new reaction products between C_3A, gypsum, and calcium chloride and the disappearance of initial products, has been studied [5, 6] and results are shown schematically in Fig. 4.1 and can be summarized as follows:

 (i) At the first contact between calcium chloride solution and cement particles, both gypsum and chloride react to form very small amounts of calcium trisulphoaluminate (ettringite) and calcium chloro-aluminate respectively.

$$C_3A + 3CaSO_4 + 32\,H_2O \rightarrow C_3A.\,3\,CaSO_4.\,32\,H_2O$$

$$C_3A + CaCl_2 + 10\,H_2O \rightarrow C_3A.3\,CaSO_4.32\,H_2O$$

 (ii) The gypsum continues to react to form ettringite whilst the calcium chloride does not react, but can continue to promote the silicate hydration.

$$C_3A + 3\,CaSO_4 + 32\,H_2O \rightarrow C_3A.3\,CaSO_4.32\,H_2O$$

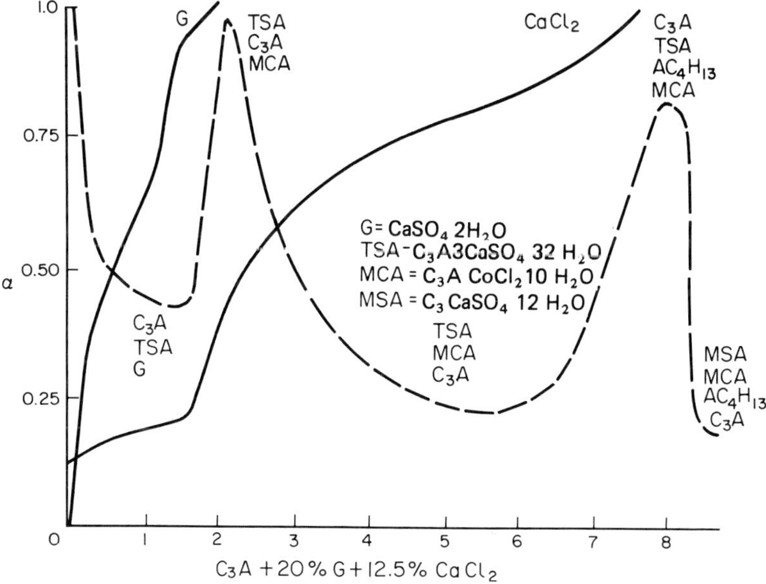

Fig. 4.1 The composition of a paste consisting of C_3A and gypsum in the presence of calcium chloride at various times (Tenoutasse).

(iii) When all the gypsum has been consumed, the calcium chloride begins to react again with the C_3A until the chloride in solution is reduced to about nil.

$$CaCl_2 + C_3A + 10\,H_2O \rightarrow C_3A.CaCl_2.10\,H_2O$$

There is some uncertainty as to the exact nature of the chloroaluminate formed and although it is likely that the monochloroaluminate is predominantly formed, it is possible that some trichloroaluminate is also present as a reaction product.

(iv) When the chloride ion has been removed, the remaining C_3A is hydrated to give $C_3A.\ Ca(OH)_2.\ 12\,H_2O$, which then converts the trisulphoaluminate to the monosulphoaluminate.

$$C_3A.\ 3\,CaSO_4.\ 32\,H_2O + 2\,C_3A.\ Ca(OH)_2.\ 12\,H_2O \rightarrow$$

$$3\,C_3A.\ CaSO_4.\ 12\,H_2O + 20\,H_2O + 2\,Ca(OH)_2$$

The final composition, therefore, consists of: $C_3A.\ Ca(OH)_2.\ 12\,H_2O$, $C_3A.\ CaCl_2.\ 10\,H_2O$, and $C_3A.\ CaSO_4.\ 12\,H_2O$ in solid solution.

(v) There is also some indication that a reaction can occur between lime and calcium chloride to form $3\ CaO.CaCl_2.16\,H_2O$ [7], particularly at low temperatures.

The reactions occurring with calcium formate, nitrate and thiosulphate have not been widely studied, although it seems likely that calcium formate and thiosulphate [6] react with C_3A in a similar manner to calcium chloride to form $C_3A.Ca(HCOO)_2.\ xH_2O$ and $C_3A.Ca(S_2O_3)_2.\ yH_2O$, respectively. In the case of calcium nitrate there is a difference in behaviour and the C_3A hydration is promoted to the cubic C_3AH_6 form.

Triethanolamine

Triethanolamine is not an effective accelerator when used alone because of its adverse effect on the resultant strength of the hardened paste. It is used as an ingredient in some admixture formulations, however, and investigations have shown that chemical interactions occur [8, 9].

(a) In the presence of TEA the reaction between C_3A and gypsum is accelerated.
(b) The subsequent conversion of the ettringite to monosulphate by reaction with C_3A is also accelerated by TEA.
(c) The formation of the hexagonal aluminate hydrate and conversion to the cubic form is accelerated by TEA.

(d) There is some evidence for the formation of a surface complex between C_2S and C_3S initial hydrates and TEA.

4.3.3 Effects on cement hydration

(a) Kinetics of reaction

Conduction calorimetric curves of Portland cement hydrated isothermally containing various quantities of triethanolamine are shown in Fig. 4.2 [8]. On initial contact with water each sample evolves heat (not shown in figure) that can be attributed to heat of wetting, hydration of free lime and reaction of C_3A with gypsum to form ettringite. The amount of heat developed increases with the quantity of TEA indicating that the C_3A + gypsum reaction is accelerated.

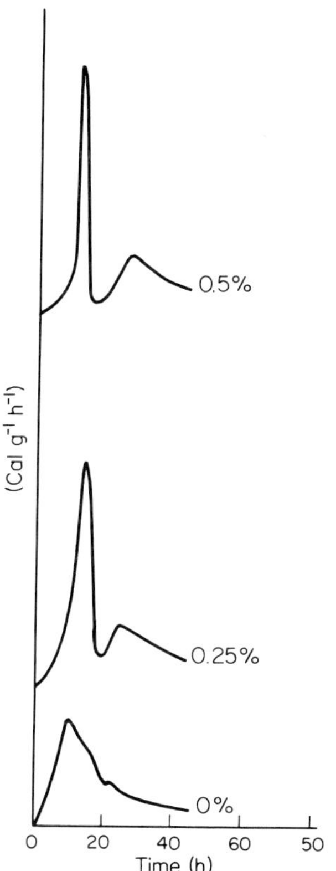

Fig. 4.2 Conduction calorimetric curves of cement hydrated in the presence of triethanolamine (Ramachandran).

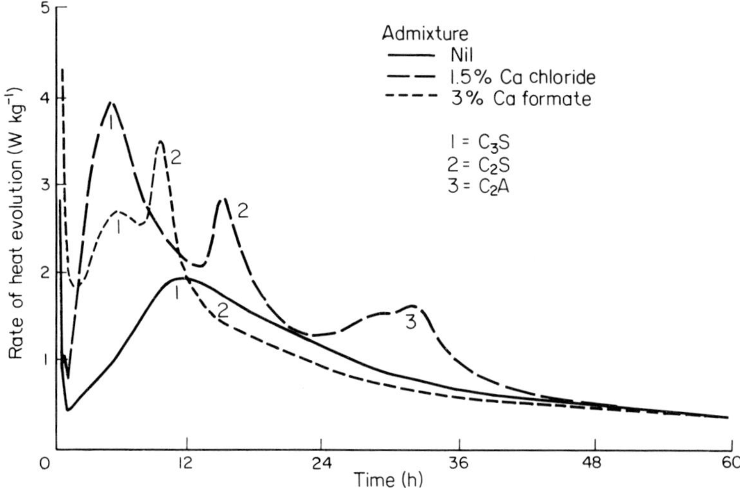

Fig. 4.3 Conduction calorimetric curves of plain and calcium chloride and formate containing pastes (Edmeades).

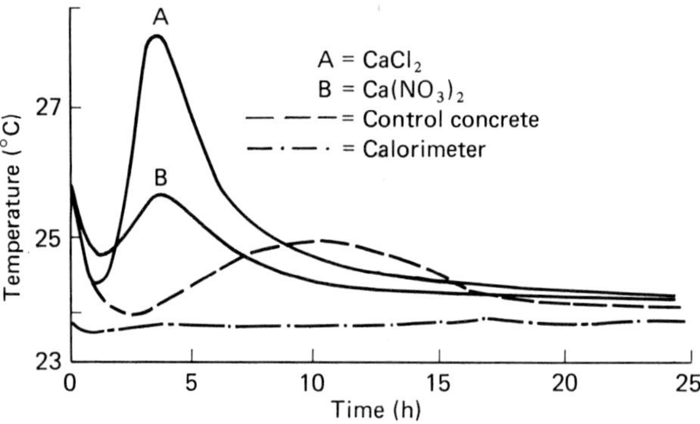

Fig. 4.4 Conduction calorimetric curves of plain and calcium chloride and nitrate containing pastes (Edwards).

The second peak, occurring after 9 to 10 h in a plain cement paste, is mainly due to C_3S hydration and in the presence of TEA is extended, suggesting that the C_3S hydration is retarded, particularly at high amounts of TEA.

Similar isothermal calorimetric curves for calcium chloride, formate [7] and nitrate [10] are given in Figs 4.3 and 4.4, which indicate that

(i) Calcium chloride is a more effective accelerator than the other materials.

(ii) All the materials accelerate reaction of the C_2S and C_3S phase hydrations, which are the main strength contributing components of Portland cement.
(iii) The C_3A phase reactions are either diminished or retarded beyond the area of the curves.

Studies of the kinetics of the C_3S hydration in the absence and presence of accelerators show that the extent or degree of hydration of the silicate phase in the presence of calcium chloride is considerably increased, right up to at least 28 days, whether measured by the quantity of lime produced [6] (Fig. 4.5), X-ray analysis [11] (Fig. 4.6), or the amount of non-evaporable water [12] (Fig. 4.7). Fig. 4.5 also shows that a small amount of TEA retards the hydration of the C_3S phase for a considerable time, and the trend of the curves would suggest that the presence of TEA will result in permanently partly hydrated C_3S.

(b) Composition and morphology of resultant hydrates

It has already been shown that the presence of accelerating admixtures produces hydration products of a different type to those from a plain cement paste, because of chemical involvement, predominantly with the C_3A phase

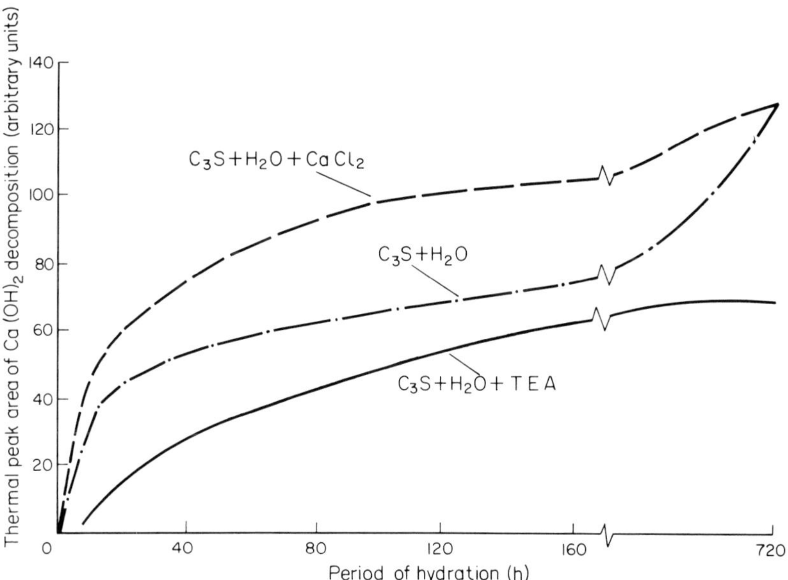

Fig. 4.5 The rate of reaction of tricalcium silicate in the presence of calcium chloride and triethanolamine, measured by lime production (Ramachandran).

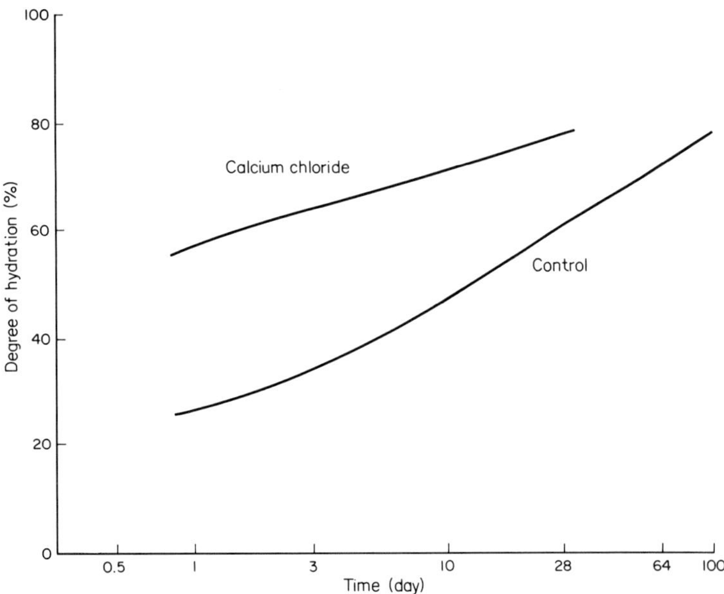

Fig. 4.6 The degree of hydration of cement pastes in the presence of calcium chloride in comparison to a plain paste; measured by X-ray analysis (Young).

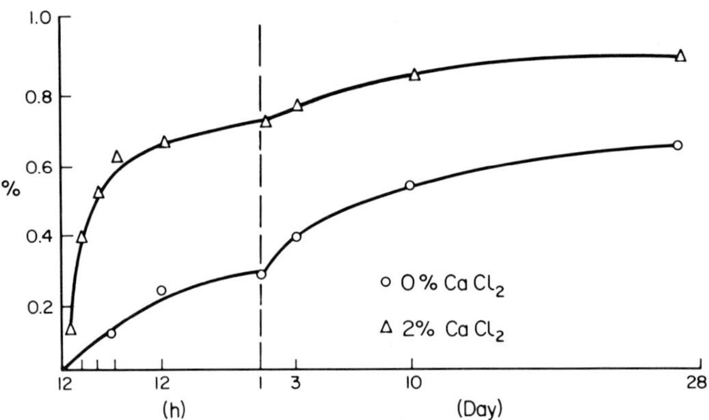

Fig. 4.7 Degree of hydration of a tricalcium silicate paste in the presence of calcium chloride, measured by non-evaporable water content (Odler).

and/or its reaction products. However, the effects on the C_2S and C_3S phases are of greater importance because they are the main strength contributing components. In addition, it is thought that the amount and type of $Ca(OH)_2$ produced, may have some effect on structural properties [13] of the hardened paste. The following observations have been made:

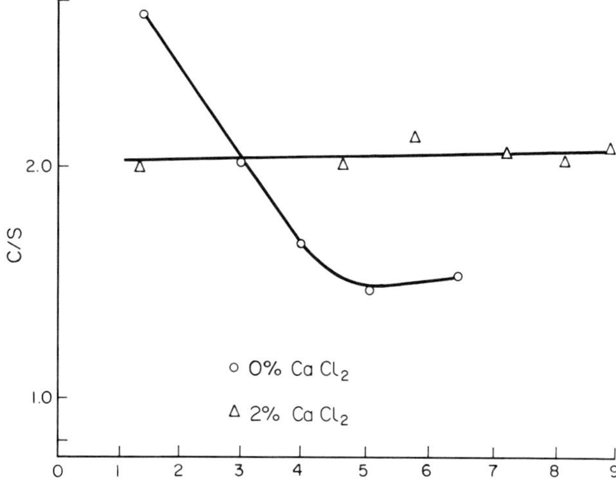

Fig. 4.8 The lime:silica ratio of hydration products of tricalcium silicate hydrating in the presence of 0 and 2% calcium chloride as a function of the degree of hydration (Odler).

(i) Calcium chloride results in a tobermorite gel with a much higher lime:silica ratio than a plain paste [12], with this ratio being approximately constant over a wide range of degree of hydration, as shown in Fig. 4.8. These data can be considered alongside Fig. 4.7 to indicate that at any degree of hydration greater than 30% (older than 1 day), the hydrated cement containing $CaCl_2$ will have a much greater lime:silica ratio than a plain cement paste. The resultant gel also has a lower specific surface.

The differences in chemical composition are accompanied by differences in the morphology of the tobermorite gel. Spicular or cigar shaped rolled sheets are formed in the normal plain hydrated cement paste, whilst in the presence of calcium chloride, thin crumpled sheets or foils are formed. It has been suggested [12] that either the high lime content or adsorbed chloride prevents rolling of the sheets.

(ii) Triethanolamine has some effect on the morphology of hydrating C_3S gels [14] where it seems that the size of the fibrous particles at 2 months is greater in the presence of TEA.

(iii) The morphology of the $Ca(OH)_2$ produced during hydration is affected by the presence of many admixtures [11, 13]. It is possible to categorize the type of morphology as in Fig.4.9. Calcium formate does not appear to affect the morphology of the $Ca(OH)_2$ produced in comparison to a plain C_3S paste which falls into group II of the classification. Calcium chloride, however, changes the form to group III.

The number of $Ca(OH)_2$ crystals formed in a given volume of paste have also been studied and, although many admixtures can have a considerable effect, neither calcium chloride nor calcium formate have a significant effect.

The peaks produced by differential thermal analysis (DTA) of hydrated

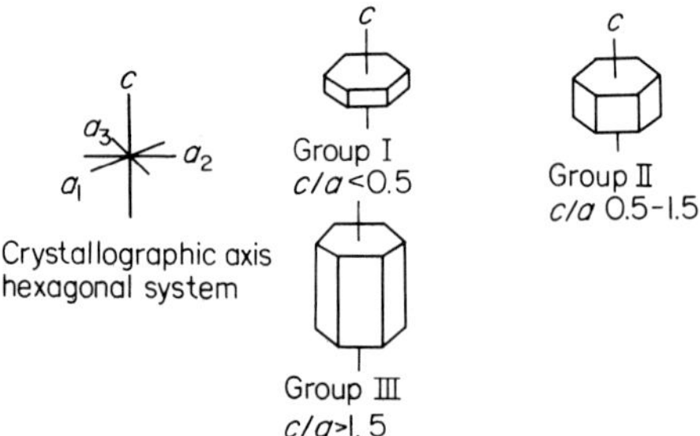

Fig. 4.9 Classification of calcium hydroxide morphology based on a *c* to *a* crystallographic axis ratio (Berger).

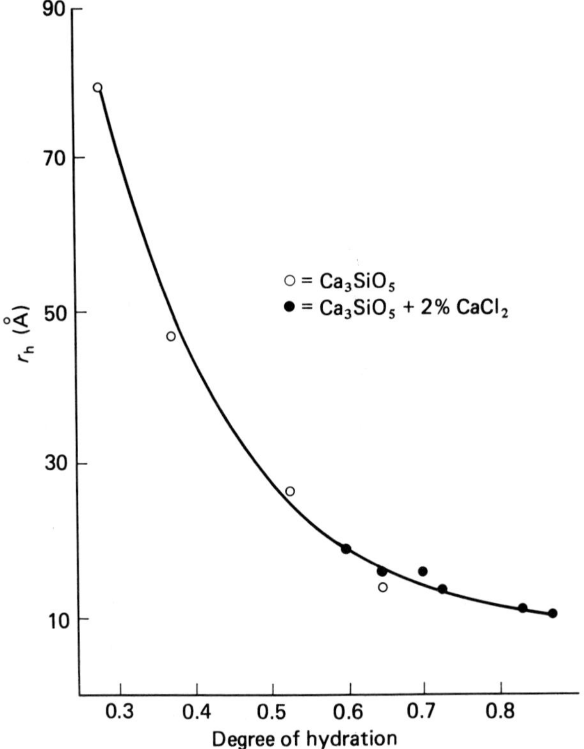

Fig. 4.10 Hydraulic radius of hydrated Ca_3SiO_5 calculated from the adsorption side of water vapour isotherms (Skalny).

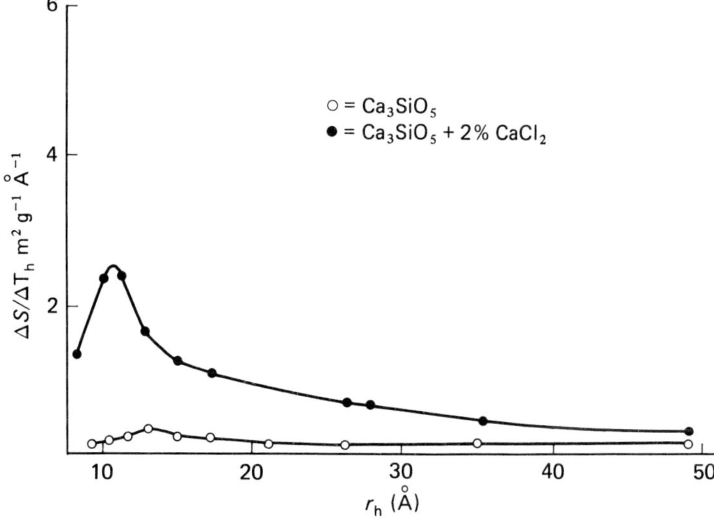

Fig. 4.11 Surface area distribution of a 28-day hydrated Ca_3SiO_5 sample measured by nitrogen adsorption (Skalny).

C_3S samples, indicate a shift in the $Ca(OH)_2$ peaks in the presence of calcium chloride [15]. It has been suggested that the $Ca(OH)_2$ may be present in a differently bonded form when calcium chloride is present.

(c) The microstructure of the hardened paste

The changes in morphology described previously for hardened pastes containing calcium chloride from the normal cigar shaped tubes or spicular to crumpled foils result in a changed gel microstructure, which can be studied using adsorption and intrusion experiments, using gases or mercury.

When the hydraulic radii of tobermorite gel pores are examined using adsorption of water vapour, very little difference is found between plain pastes and pastes containing calcium chloride [16] as shown in Fig. 4.10. However, nitrogen adsorption reveals considerable differences, as shown in Fig. 4.11, in terms of surface area distribution of a 28-day hydrated C_3S specimen [16], or of specific surface area as a function of hydration time [17] in Fig. 4.12.

This difference between H_2O and N_2 adsorption data has been attributed to either the accessibility of water to interlayer spaces in the tobermorite gel, or to the presence of 'ink bottle' pores with narrow necks and wide bodies. The considerable increase (4 to 5 times) in pore surface and pore volume available to nitrogen in pastes containing calcium chloride suggests that the crumpled pore type of morphology is more open than the spicular type [16].

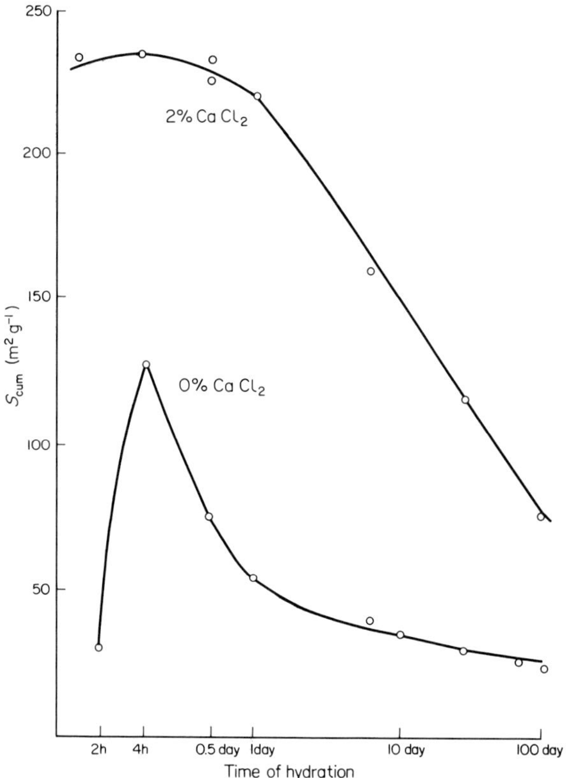

Fig. 4.12 The specific surface area of plain and $CaCl_2$ containing pastes as a function of hydration time (Collepardi).

The type of data produced by H_2O and N_2 adsorption is relevant to gel pores having radii up to about 50 Å, but larger pores and capillaries exist in the hardened cement paste and probably are more significant in determining the porosity or permeability of the hardened paste in concrete to gases and liquids.

Mercury intrusion data for larger pores of 6.5×10^{-3} to 10 μm are shown in Figs 4.13 and 4.14 at various hydration times and at equal hydration, respectively [18]. It can be seen that the total intrusion at the limit of the instrument (6.5×10^{-3} μm) is less for chloride containing pastes at all ages, but when considered at equal degrees of hydration, a higher proportion of coarse pores is formed.

4.3.4 Mechanism of action

The way in which salts such as calcium chloride and calcium formate operate is not fully understood, but it is clear that the mechanism involves an

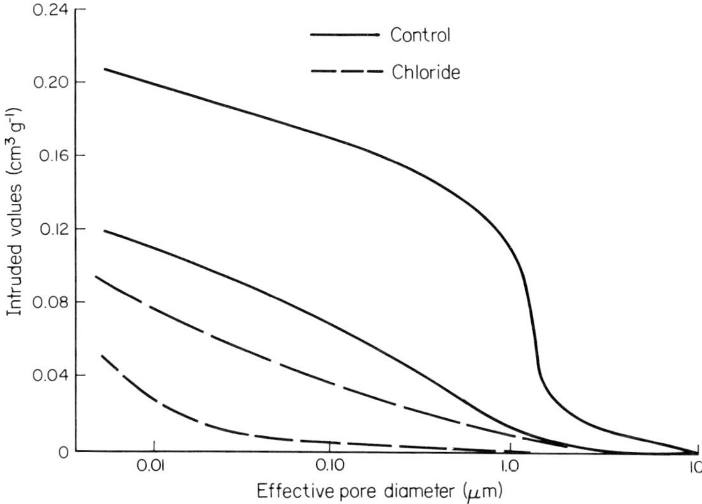

Fig. 4.13 Mercury intrusion porosimetry curves of C_3S pastes showing differences in capillary porosity distribution.

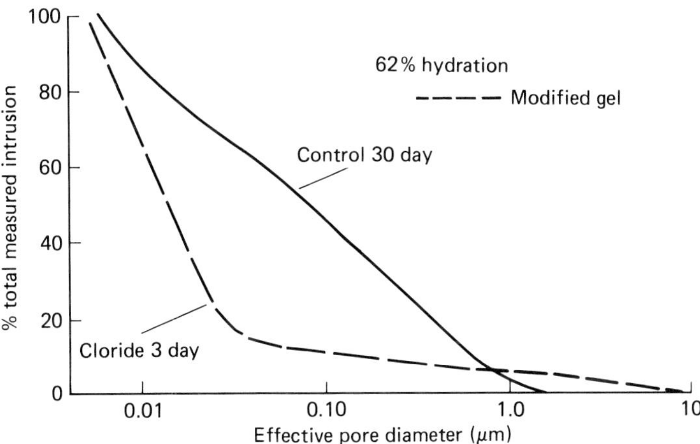

Fig. 4.14 Porosity distribution curves at equal hydration with intruded volume plotted as a percentage of total intrusion (Young).

acceleration of the C_2S and C_3S hydration. It has been proposed [19] that the inital products of cement hydration form a sort of 'membrane' which acts as a restraint to the diffusion process which in turn leads to the 'dormancy period'. It seems likely that the chloride ion, by virtue of its small size and high mobility, is able more easily to penetrate the pores of the restraining layer allowing the diffusion process to proceed more rapidly. The resultant tobermorite gel has a higher lime:silica ratio and a more open, accessible structure, based on a crumpled foil morphology rather than the usual spicular. The considerable reaction with, and modification to, the C_3A hydration is not relevant to the acceleration process.

Triethanolamine accelerates the reactions associated with the C_3A hydration, but retards, possibly permanently, the C_3S hydration, which is reflected in reduced ultimate strength.

4.4 The effects of accelerators on the properties of plastic concrete

Accelerating admixtures based on calcium chloride or calcium formate have no significant effect on the workability, air content, mix stability, or water–cement ratio of concretes into which they are incorporated. The only properties of plastic concrete which are affected are the heat evolution and setting time.

4.4.1 Effect on heat evolution

The heat evolution of concrete mixes containing no admixture, 1.5% calcium chloride and 3.0% calcium formate is shown in Fig. 4.15. The heat evolution of calcium chloride and calcium formate are approximately equal at 24 h, which is reflected in similar compressive strengths at this age of 10.0 and 12.5 N mm^{-2}, respectively, in comparison to the plain concrete mix,

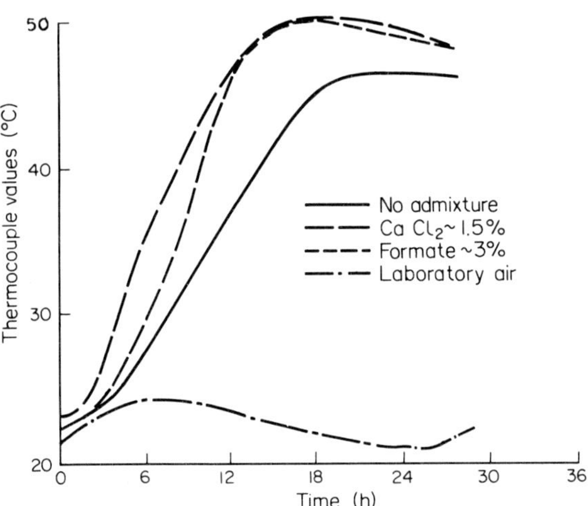

Fig. 4.15 Heat evolution from insulated concrete cubes containing calcium chloride and formate.

Table 4.1 The effect of calcium chloride and formate on mortar stiffening times

Accelerating admixture type	Stiffening time (h) for penetration resistance of 0.5 N mm^{-2}	3.5 Nmm^{-2}
None (control)	3¼	5
3.2% calcium chloride	1	1¾
2.0% calcium formate	2¼	3¼

which showed a maximum temperature rise some 4 °C lower and had a 24 h compressive strength of 5.5 mm^{-2}. Limited tests [18] on concrete mixes containing the pozzolan material, fly ash, indicate that increased heat evolution is due to the cement only, suggesting that calcium chloride may not influence the pozzolanic reaction.

4.4.2 Effect on setting time

Accelerating admixtures reduce both the initial and final setting time of mortar sieved from concrete mixes, determined by ASTM C 403:68 or BS

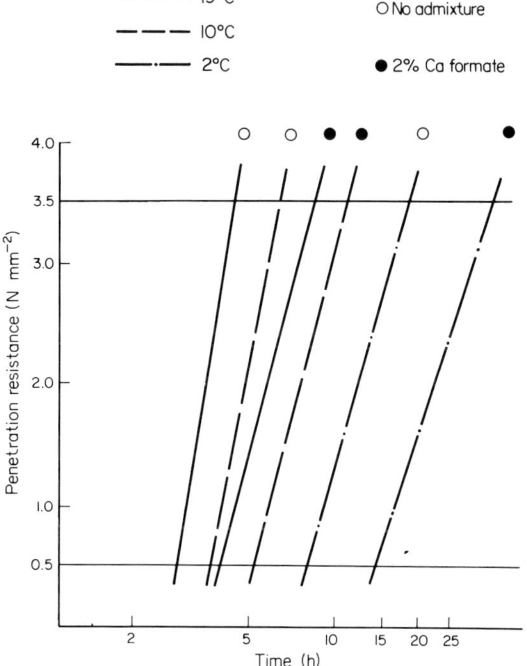

Fig. 4.16 Stiffening times of concrete with and without calcium formate accelerator (to BS 5075) (Edmeades).

5075 (1975). Typical results are given in Table 4.1 for a 300 kg m^{-3} cement content mix with a compacting factor of 0.85 ± 0.02 [20] at normal ambient temperature.

The initial and final setting times are also reduced at lower temperatures, as shown in Fig. 4.16 for a calcium formate based material [7].

4.5 The effects of accelerators on the properties of hardened concrete

4.5.1 Structural design parameters

(a) Compressive strength

The compressive strength of concrete containing an admixture of the calcium chloride type will have a higher 28-day compressive strength than a plain concrete mix having the same mix composition. This is true of concretes cured over a range of temperatures [18] as shown in Fig. 4.17. The difference is greater at lower temperatures and Table 4.2 summarizes the increases at 28 days that can be expected; these are largely independent of cement content.

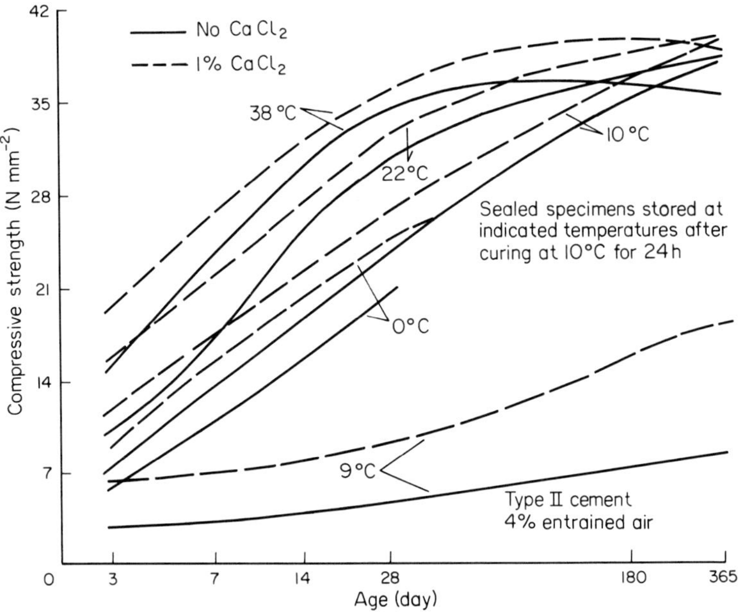

Fig. 4.17 1% $CaCl_2$ increases the compressive strength of concrete stored at various temperatures (Shideler).

Table 4.2 The increase in compressive strength of concrete containing calcium chloride is greater at lower temperatures

Curing temperature (°C)	% increase in compressive strength at 28 days over plain concrete
−10	90
0	25
10	16
20	12
40	7

(b) Flexural and tensile strength and modulus of elasticity

The data available on the effect of accelerators on the flexural and tensile strengths and the modulus of elasticity are limited [18, 21, 22] and are relevant only to calcium chloride and triethanolamine. The data generally point to either no effect or a slight reduction in all three properties. Also, in view of the increase in compressive strength, it is reasonable to assume that for a given compressive strength, the presence of calcium chloride or triethanolamine will reduce the flexural and tensile strengths and the modulus of elasticity.

4.5.2 Durability aspects

(a) Permeability

The permeability and porosity of concrete containing calcium chloride in relation to a plain concrete depends on two conflicting variables:

(i) The degree of hydration of the concrete which in the case of the calcium chloride containing concrete will initially be considerably increased and the larger volume of hydration products will lead to a reduced permeability.

(ii) The adverse effect that calcium chloride has on the capillary porosity distribution. At later ages (after perhaps 1 year) the degree of hydration in both calcium chloride containing concrete and a plain concrete will be similar and, under these circumstances, the concrete will be more porous and allow easier access to aggressive gases and liquids. Pore volume distributions of cement pastes at an advanced state of hydration of more than 90% [23] are shown in Fig. 4.18 where the greater porosity is indicated and the data can be used to calculate an increase in average hydraulic radius from 28 Å in the chloride free sample, to 108 Å in the chloride containing sample.

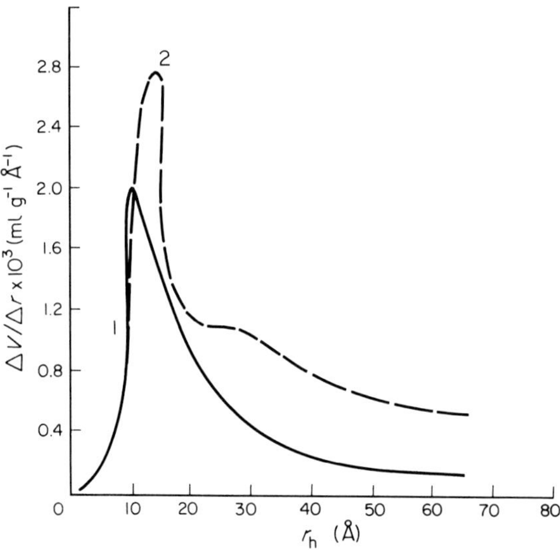

Fig. 4.18 Pore volume distribution of cement pastes from nitrogen adsorption. Curve 1 = cement with no admixture; curve 2 = cement paste and 2% $CaCl_2$ (Gouda).

(b) Resistance to aggressive environment

The resistance of concrete containing calcium chloride to attack by aqueous sulphate is reduced [18, 22, 24]. A comprehensive study of concrete over a 5 year period [20] using various cements and cement content stored in high sulphate containing water, gave the results shown in Figs 4.19 to 4.24, for which the following conclusions can be reached:

(i) Calcium chloride at 2% addition level contributed to a lowering of sulphate resistance in almost all cases.
(ii) The higher C_3A type I and II (American classification) cements were more adversely affected than the low C_3A type V cement.
(iii) The low cement content mixes were drastically affected with regard to both expansion and compressive strength reduction.
(iv) Only mixes containing 426 kg m^{-3} low C_3A cement showed unaffected resistance to sulphate attack over a 5 year period.

(c) Resistance to freeze–thaw conditions

Concrete containing calcium chloride develops strength more rapidly and, therefore, has a greater resistance to damage by freezing at an early age, as shown in Fig. 4.25. There is some indication, however, that at later ages, the more mature concrete is less resistant to freeze–thaw cycling [18, 24].

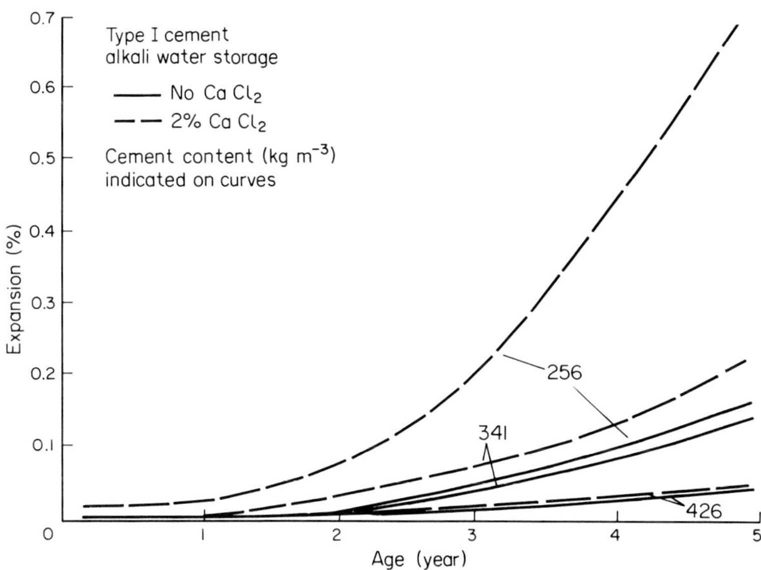

Fig. 4.19 $CaCl_2$ decreases the resistance of concrete to sulphate attack (Shideler).

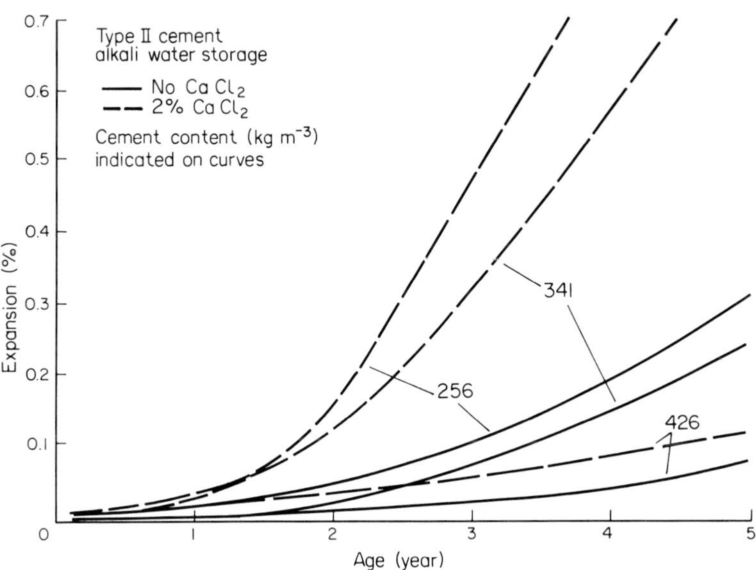

Fig. 4.20 $CaCl_2$ decreases the resistance to attack, particularly in lean concrete (Shideler).

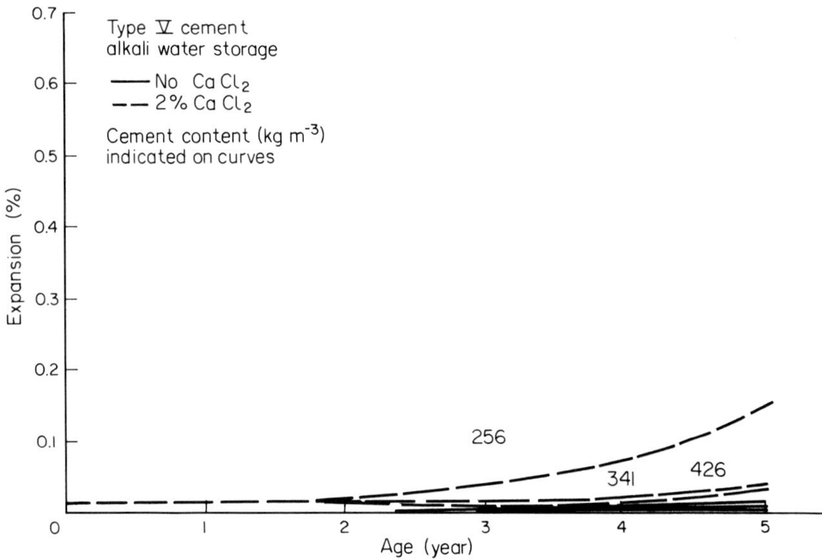

Fig. 4.21 $CaCl_2$ is least detrimental to sulphate resistance with type V cement (low C_3A) (Shideler).

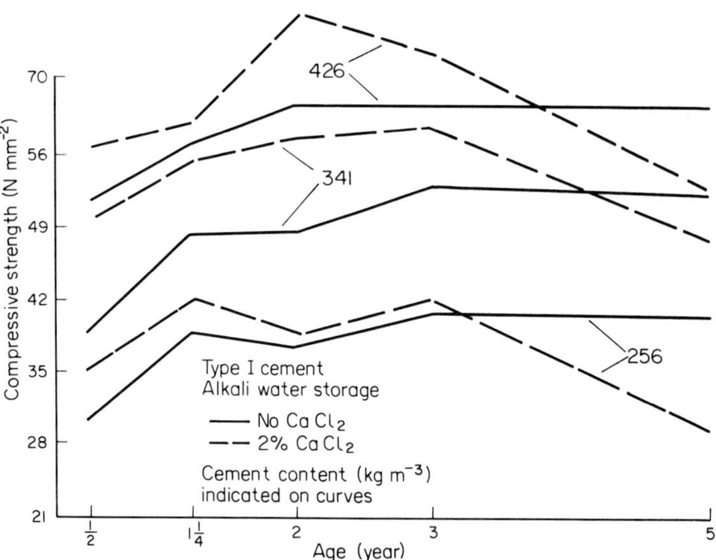

Fig. 4.22 $CaCl_2$ reduces the compressive strength of concrete under sulphate attack (Shideler).

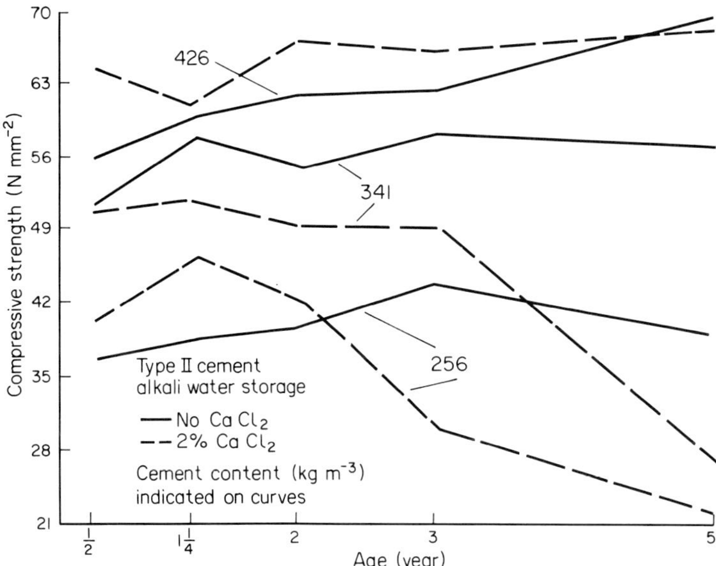

Fig. 4.23 $CaCl_2$ reduces the compressive strength of lean concrete under sulphate attack (Shideler).

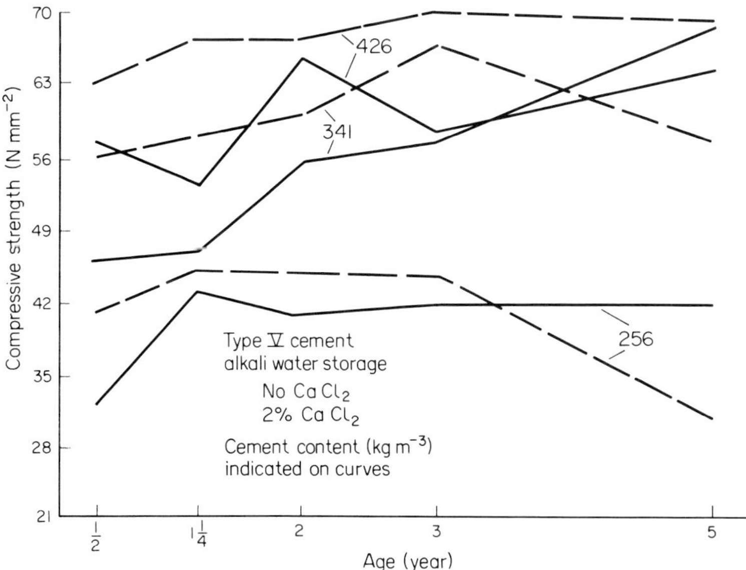

Fig. 4.24 $CaCl_2$ has little effect on sulphate resistance with type V cement (low C_3A) (Shideler).

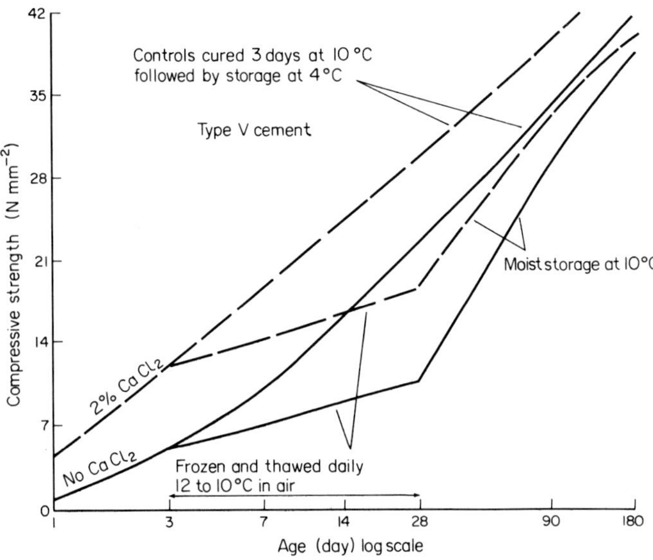

Fig. 4.25 $CaCl_2$ increases the strength of concrete frozen at early ages (Shideler).

(d) Protection of steel reinforcement

The effect that accelerators have on the role of concrete in providing protection against the corrosion of steel reinforcement has been the subject of several investigations and considerable controversy. Many studies have shown that the factors below are relevant to the discussion:

(i) The porosity of the concrete at an advanced state of maturity is increased in the presence of calcium chloride and, therefore, will allow a greater opportunity for air and moisture to come into contact with the steel reinforcement, encouraging corrosive effects. In practice, with reinforcement cover meeting the relevant codes of practice, this effect is regarded as of minimal significance.

(ii) The presence of calcium chloride at concentrations greater than about 1.5% by weight of cement can lead to breakdown of the passive layer of γ-Fe_2O_3 normally present at the steel/concrete interface, rendering the reinforcement more susceptible to corrosion. This does not occur with calcium formate based materials and can be shown by recording the potential of a steel/concrete electrode relative to that of a saturated calomel electrode at a constant current density as a function of time. A typical circuit [25] is shown in Fig. 4.26, and results for calcium chloride and a calcium formate based material [26] are shown in Fig. 4.27 where

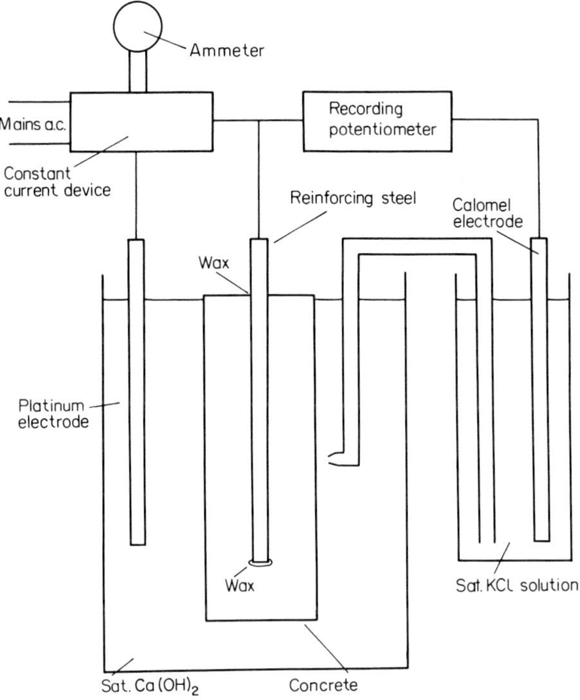

Fig. 4.26 Typical circuit for recording the potential of a steel/concrete electrode relative to that of a saturated calomel electrode (Gouda).

the breakdown of the passive layer is seen for 3% calcium chloride. There is some evidence that the presence of chlorides not only renders the steel more liable to corrosion attack, but also alters the crystallographic nature of the initial corrosion products from the normal orthorhombic form of Fe_2O_3 to a tetragonal form [27].

(iii) Visual inspection of steel surfaces which have been embedded in concrete and then broken open, indicates that the chloride ion has a chemical action on the steel which, combined with the local electrochemical cell action imposed by the distribution of narrow and wide pores over the steel surface and local concentrations of air over the surface, causes a 'pitting' type of corrosion of the steel surface. Table 4.3 gives results for the corrosion of narrow steel wires embedded in concrete for periods of 2 years in terms of the maximum depth of pits, the mean depth of the three deepest pits and the average number of pits for each group of embedded wires [28]. It can be seen that in concretes containing ordinary Portland cement, only slight pitting occurs when

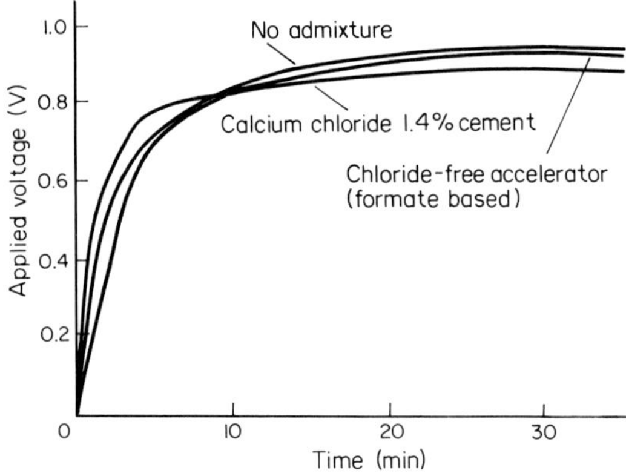

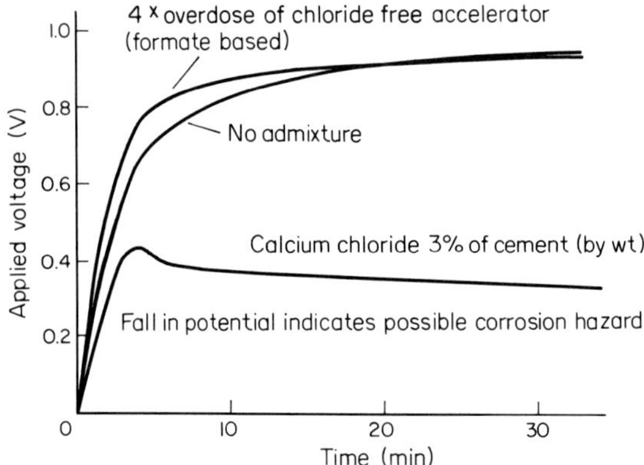

Fig. 4.27 Potential–time curves for plain concretes and concretes and concretes containing calcium chloride and calcium formate based accelerators (McCurrich).

2% by weight of calcium chloride is included, but severe pitting occurs with a 5% addition. In mixes containing sulphate resistant cement, pitting is considerable even at 2% addition level.

(iv) Any corrosive attack of normal diameter reinforcing steel is not usually enough to cause significant reductions in the tensile strength of the steel in the early stages. The problem lies in the spalling of concrete away from the reinforcement because of the expansive nature of the products of corrosion. The resultant exposure of the reinforcement allows further corrosion which could reduce the tensile strength of the steel sufficiently to cause structural failure.

Table 4.3 Effect of $CaCl_2$ on corrosion of high tensile steel wires in normally cured prestressed concrete specimens stored outdoors

Cement	Flake $CaCl_2$ (%)	Pitting values* on wires cleaned in 'Clarke's solution'		
		6 month	12 month	24 month
Ordinary Portland	0	< 2.5 : — : —	2.5 : — : —	2.5 : 2.5 : —
	2	5.1 : 5.1 : —	17.8 : 17.8 : 2.5	17.8 : 10.2 : 2.5
	5	50.8 : 38.1 : 22.9	99.1 : 73.7 : 15.2	127.0 : 121.9 : 25.4
Sulphate resisting Portland	2	86.4 : 58.4 : 5.1	76.2 : 71.1 : 10.2	101.6 : 81.3 : 17.8
	5	53.3 : 48.3 : 35.6	116.8 : 101.6 : 17.8 (1)	127.0 : 109.2 : 22.9 (3)

*Pitting values are expressed as maximum depth in mm $\times 10^{-2}$: mean depth in mm $\times 10^{-2}$ of the three deepest pits : average number of pits per wire for the group of eight wires in each concrete specimen. Figures in brackets refers to number of breaks in wires due to corrosion. Wires were prestressed and 2.03 mm in diameter.

(v) The presence of an excessive number of voids in the concrete [29] or porosity due to poor compaction or a deficiency of fine aggregate [30] leads to an increase in the amount of reinforcement corrosion.

(vi) The bond between concrete and reinforcing steel appears to be greater in concrete containing calcium chloride and increases with calcium chloride concentration as shown in Fig. 4.28 [31] which may be a function of the higher compressive strength and shrinkage of the accelerator containing concrete.

The general conclusion can be reached that in well designed properly compacted concrete, the addition of up to 1.5% by weight of cement of calcium chloride can be used without any significant detrimental effect on embedded ferrous metals.

Excessive dosages of calcium chloride, particularly in concrete of high porosity, can lead to accelerated corrosion with subsequent spalling and possible structural failure. Recent changes in codes of practice preventing the use of calcium chloride in concrete containing embedded metal are based on the difficulties of controlling addition levels under site conditions and the possibilty of porous or poorly compacted concrete being produced.

There is very limited information available on the effect of calcium formate but certainly the passive layer at the concrete/steel interface does not appear to be attacked.

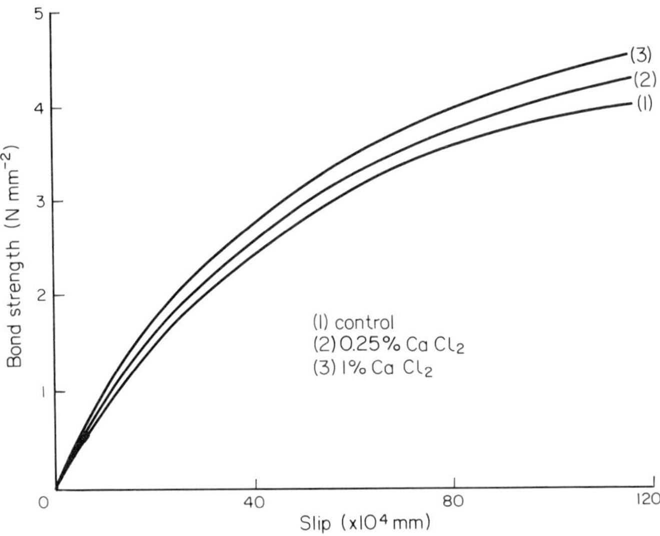

Fig. 4.28 Bond strength against slip for plain concrete and concrete containing calcium chloride (Kondo).

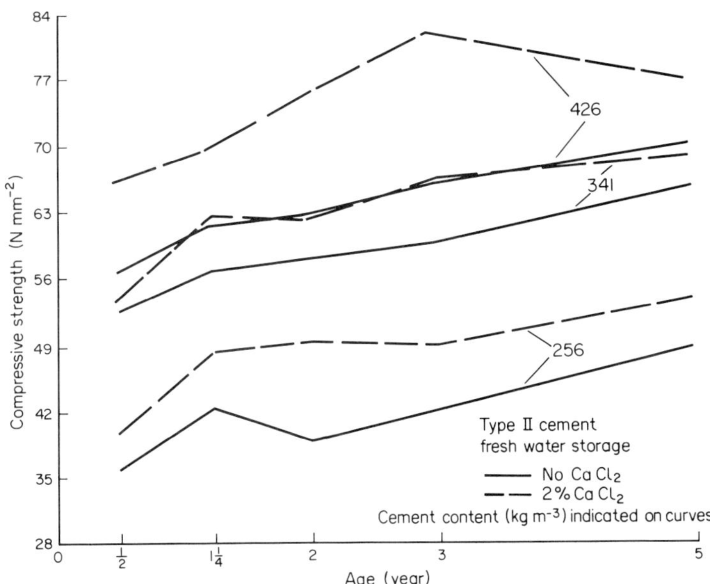

Fig. 4.29 Calcium chloride does not adversely affect the strength of concrete up to the five years studied (Shideler).

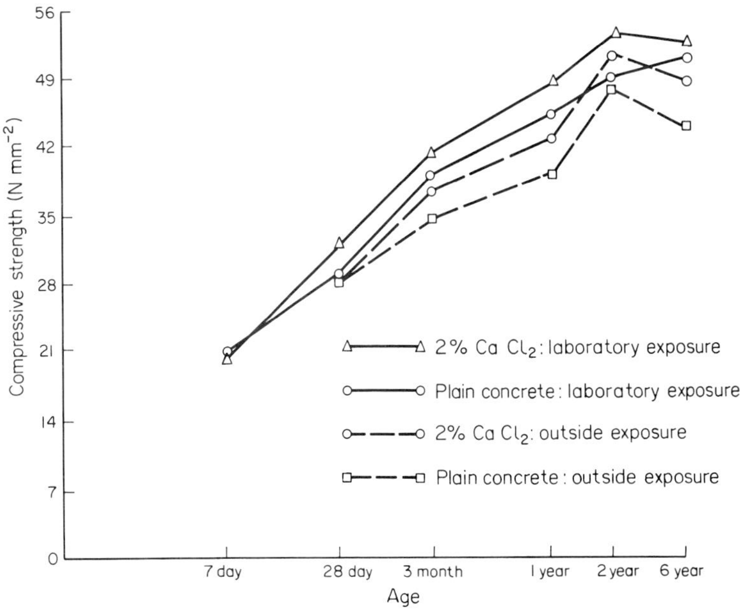

Fig. 4.30 Calcium chloride does not affect the long term strength development of dense concrete cured under laboratory conditions, or with outside exposure (Blenkinsop).

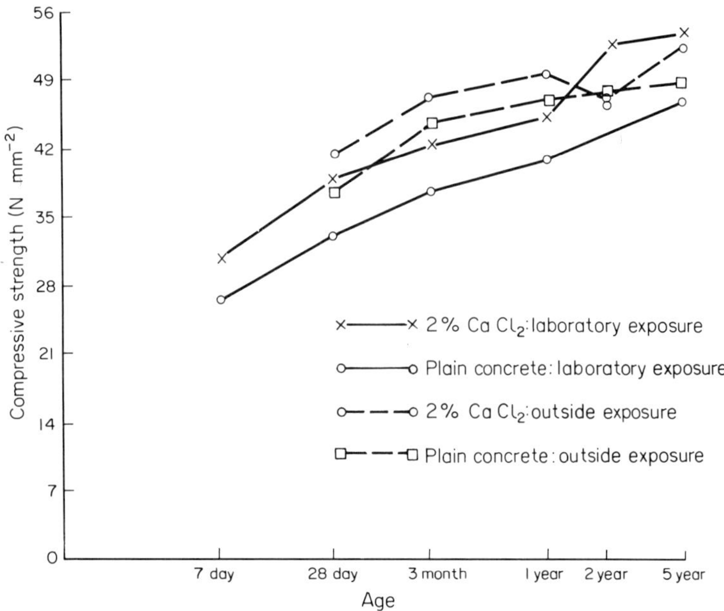

Fig. 4.31 Calcium chloride does not affect the long term strength development of porous concrete cured under laboratory conditions, or with outside exposure (Blenkinsop).

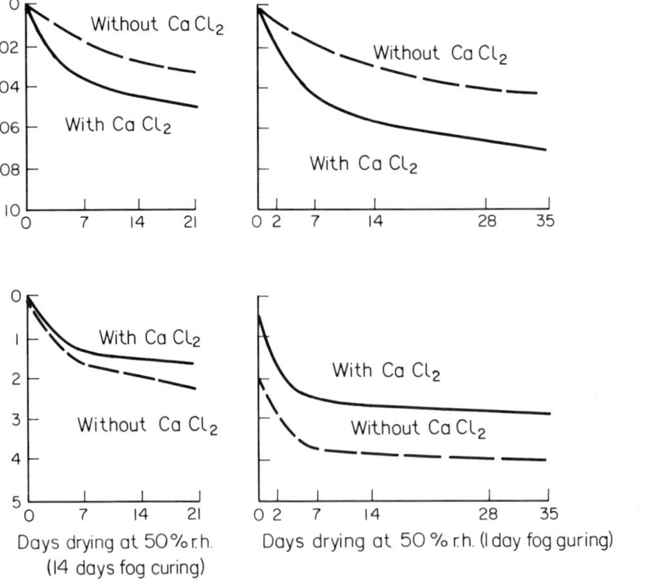

Fig. 4.32 $CaCl_2$ increases drying shrinkage of concrete, although the moisture loss is decreased (Shideler).

Table 4.4 Drying shrinkage of concretes containing calcium chloride, triethanolamine and calcium formate

Admixture	Concn. % by wt of cement	W/c ratio	Cement content (kg m^{-3})	Slump (mm)	% drying shrinkage at (day)					
					7	14	28	56	84	168
None*	—	0.60	305	65	0.013	0.020	0.031	0.043	0.049	0.056
Calcium chloride*	2.0	0.60	305	60	0.023	0.033	0.045	0.056	0.062	0.068
Triethanolamine*	0.033	0.60	305	65	0.016	0.029	0.040	0.053	0.056	0.064
None†	—	0.60	305	65	0.016	0.018	0.027	0.039	0.045	0.052
Calcium formate†	3.00	0.60	305	65	0.015	0.025	0.036	0.048	0.054	0.056

*Average of mixes containing five different cements.
†One mix only.

(e) Compressive strength development

Although the main purpose of using accelerators is to obtain high early strength, the increased strength has been found [18, 30] to continue for a period of several years when addition levels of up to 2% of calcium chloride by weight of cement are used, and Figs 4.29 to 4.31 illustrate this effect for specimens stored for 5 years under water (Fig. 4.29), in laboratory atmosphere (Fig. 4.30), and under outdoor exposure (Fig. 4.31). However, higher percentages of calcium chloride (greater than 4%) will lead to reduced strengths in comparison to a plain cement at periods greater than about 1 year.

(f) Volume deformations

(i) *Shrinkage.* The drying shrinkage of concrete containing calcium chloride is increased in comparison to plain concrete, even though the amount of moisture lost is less [18]. This is illustrated in Fig. 4.32 and it is thought that the reduced moisture loss will be due to the more advanced state of hydration in the specimens containing calcium chloride. The increased shrinkage must, therefore, be a characteristic of the type of cement hydration products formed.

Under saturated conditions, such as total water immersion, the amount of expansion of the concrete is reduced when calcium chloride is present.

There are only limited data available on the effect of other accelerating admixtures, although one comparative study [32] suggests that calcium formate and triethanolamine also increase the drying shrinkage of concrete

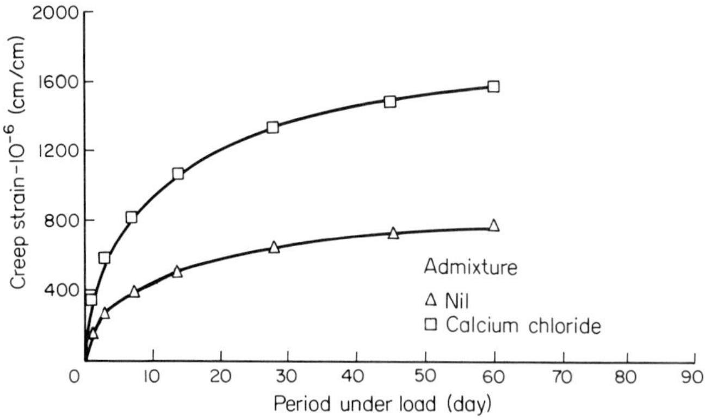

Fig. 4.33 The creep of plain concrete and concrete containing calcium chloride (Hope).

into which they are incorporated and some of this information is summarized in Table 4.4 in comparison to plain concretes and concretes containing calcium chloride.

(ii) *Creep*. The creep of concrete under drying conditions is increased by the presence of calcium chloride in the mix [21] as shown in Fig. 4.33 for 1.5% calcium chloride by weight of cement.

References

1 Rosenburg, A.M. (1974). *ACI Journal,* **71,** 1261–8.
2 US Patent (1974). 3801 338.
3 Murakami, K. (1968). *Proceedings of the 5th International Symposium on the Chemistry of Cement,* Tokyo, Supplement paper II – 2.
4 Wolhuter, C.W. (1974). *Proceedings of the 1st Australian Conference on Engineering Materials,* University of New South Wales, 29–43
5 Tenoutasse, N. (1968). *Proceedings of the 5th International Symposium on the Chemistry of Cement,* Tokyo, 372–8
6 Ramachandran, V.S. (1972). *Thermochemica Acta,* **4,** 343–66
7 Edmeades, R.M. (1976). Private communication.
8 Ramachandran, V.S. (1976). *Cement and Concrete Research,* **6,** 623–32.
9 Ramachandran, V.S. (1973). *Cement and Concrete Research,* **3,** 41–54.
10 Edwards, G.C. (1966). *Journal of Applied Chemistry,* **16,** 166–8.
11 Young, J.F. (1973). *Cement and Concrete Research,* **3,** 689–700.
12 Odler, I. (1971). *Journal of the American Ceramics Society,* **54,** 362–3
13 Berger, R.L. (1972). *Cement and Concrete Research,* **2,** 43–55.
14 Ciach, T.D. (1971). *Cement and Concrete Research,* **1,** 159–76.
15 Ben-Dor, L. (1975). *Journal of the American Ceramics Society,* **58,** 87–9.
16 Skalny, J. (1971). *Journal of Colloid and Interface Science,* **35,** 434–40.
17 Collepardi, M. (1972). *Cement and Concrete Research,* **2,** 57–65.
18 Shideler, J.J. (1952). *ACI Journal,* **48,** 537–59.
19 Double, D.D. *et al.* (1978). *Proc. Roy. Soc.* **A 360,** 435.
20 Fletcher, K.E. (1971). *Concrete,* **5,** 175–9.
21 Hope, B.B. (1971). *ACI Journal,* **68,** 361–5.
22 Monofore, G.E. (1960). *ACI Journal,* **56,** 491–515.
23 Baker, C.A. (1966). Humes Technical Bulletin, No. 4.
24 Gouda, U.K. (1973). *Journal of Colloid and Interface Science,* **43,** 293–302.
25 Ramachandran, V.S. (1975). *Precast Concrete,* **6,** 149–151.
26 Gouda, U.K. (1966). *British Corrosion Journal,* **1,** 138–142.
27 McCurrich, L.H. (1975). *Proceedings of the Conference on Ready-Mixed Concrete,* Dundee.
28 Murat, M. *et al.* (1974). *Cement and Concrete Research,* **4,** 945–52.
29 Roberts, M.H. (1962). *Magazine of Concrete Research,* **14,** 143–54.
30 Blenkinsop, J.C. (1963). *Magazine of Concrete Research,* **15,** 33–8.
31 Kondo, Y. *et al.* (1959). *ACI Journal,* **55,** 299–312.
32 Bruere, G.M. *et al.* (1971). *CSIRO (Australia) Division of Applied Mineralogy,* Technical Paper No. 1.

5 Applications of admixtures

5.1 Introduction

Most of the concrete placed in North America and the Middle East, and an increasing proportion of that used in Europe and other parts of the world, whether precast, site batched or ready-mix, contains an admixture.

Important prerequisites for the successful performance of an admixture are proper selection and the use of an appropriate method of storage and batching. Information is presented on both the role of an admixture under a given set of conditions and methods used to ensure that the admixture is accurately dispensed into the mix. The use of admixtures in both general and special concreting techniques in the three main construction fields of ready-mix, precast and site batched concrete is discussed and operating procedures associated with the batching and handling of admixtures described. An overview of the economy that results from the use of the various admixtures is also presented.

Admixtures are generally used for the following reasons:

(i) To enable the concrete to meet requirements of job specifications, namely, permitted maximum water–cement ratio, minimum early and ultimate strengths, and retention of workability when the available raw materials are of poor quality.
(ii) To provide modifications which improve the quality of plastic concrete, e.g. increased workability or water reduction at given consistencies, improved finishing qualities, controlled bleeding and segregation.
(iii) To reduce the cost of concreting operations by effecting a reduction of the overall cost of concrete ingredients, permitting rapid mould turn over and ease of placing and finishing.

(iv) To improve the quality of hardened concrete such as increased early and long term strengths and modulus of elasticity, decreased permeability and absorption, increased abrasion resistance and increased bond with reinforcement.

5.2 Air-entraining admixtures

Air entrainment has provided great benefit in many applications, but nowhere more dramatically than in situations where the structure is exposed to freeze–thaw conditions. Susceptibility to damage by freeze–thaw cycles

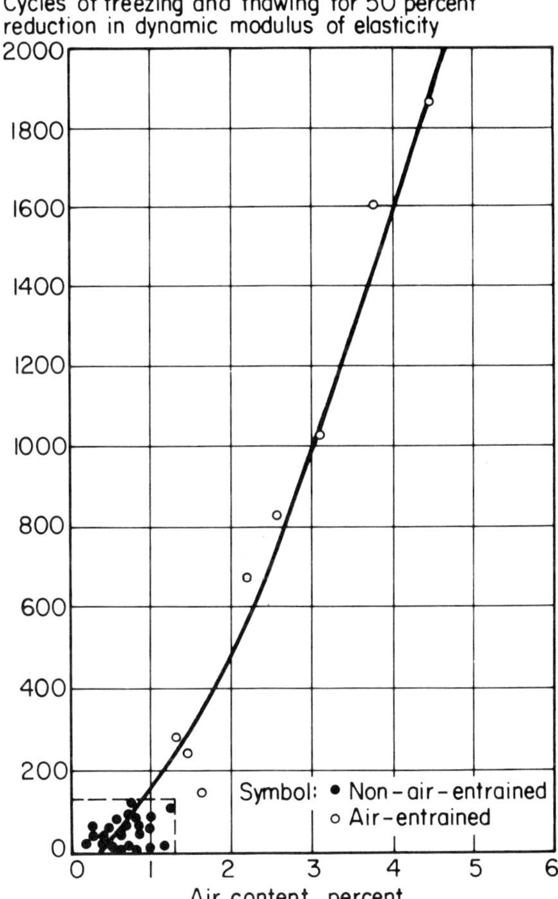

Fig. 5.1 Effect of entrained air on the resistance to freezing and thawing in laboratory tests.
(Courtesy, Portland Cement Association, Skokie, Illinois, USA) [2]

Fig. 5.2 Effect of freeze–thaw cycles on concrete with and without air entrainment. (a) Concrete without air-entrainment local aggregate. This sample failed at 100 cycles of freezing and thawing. Expansion 0.6%, weight loss 25%. Since 25% loss is considered equivalent to failure, further cycles of freezing and thawing were discountinued on the above sample.
(b) Air-entrained concrete 4% air local aggregate. After 300 cycles of freezing and thawing, expansion was 0.2% and weight loss 4%. The above sample shows the direct advantages that occur through use of air entrainment in concrete.
(Photograph – Courtesy, Protex Industries Inc. Denver, Colorado, USA)

exists where moisture in concrete reaches the critical saturation point, with the severity of the deterioration depending on the degree of exposure.

In the US and Canada not only is the concrete exposed to severe freeze–thaw cycles, but it is also often subjected to the damaging effects of ice removal salts. Highway bridges often span water courses and abutments and piers are situated in areas where wet and dry conditions alternate. Certain parts of hydraulic structures such as dams, spillways, tunnel inlets and canal structures are exposed to fluctuating water levels or spray. Concrete in these structures is often saturated and quite vulnerable to freeze–thaw damage [1]. Consequently, a large portion of the concrete placed in North America is air entrained.

In the United Kingdom the application of air-entrained concrete is largely confined to the production of concrete carriageways, aircraft runways, lightweight concrete and precast concrete cladding panels, otherwise concrete mixes are not usually designed for frost resistance.

In order to provide adequate protection against freeze–thaw action most North American specifications for durable concrete always include the requirement of low water–cement ratio and the proper amount of entrained air consistent with the maximum aggregate size used. Figs 5.1 [2] and 5.2 show the effect of entrained air on the resistance of concrete to freezing and thawing conditions.

The air-entraining agents used can be conventional neutralized wood resins, or the more recent types which include synthetic detergents (alkyl aryl sulphonates), salts of petroleum acids (sodium salts of naphthenic acid), salts of fatty acids, and organic salts.

Achievement of the benefits of air entrainment in a consistent manner requires close control of the proportioning and batching of the concrete and a good understanding of the factors which affect air-entraining characteristics. It is important that the concrete producer ensures that the properties of the concrete materials, mix proportioning and all aspects of the mixing, handling and placing procedures are maintained as constant as possible. Some control measures that should be used particularly in ready-mix concrete include the following [3]:

(i) Air-entraining admixtures of proven performance should be used with an accurate dispensing system.
(ii) Special attention should be paid to proper batching technique and maintenance of fins in the mixers to assure formation of a proper air void system.
(iii) Concrete should arrive at the job site with the proper slump. Excess water added at the job site (re-tempering) could cause the air bubbles to cluster thereby reducing durability.
(iv) Air content should be checked both at the plant and at the job site.
(v) Air loss from plant to job site should be established and admixture dosage requirements adjusted at the plant to compensate.
(vi) Other parameters of the air void system in the hardened concrete should also be determined to confirm that the air void spacing factor is within the limits specified in the standards.

It has been shown that the air content of a concrete mix is usually indicative of the adequacy of the air void system when the admixture used meets pertinent specifications of the national standards (ASTM C-260, CSA A.266.2 and BS 5075).

The recent increased use of pozzolanic and supplementary cementing materials such as fly ash and slag in the ready-mixed industry requires close control of air-entrained concrete. This is due to the difficulty of entraining the required air content when these materials are present in the concrete mix.

For concrete mixes containing fly ash with carbon content less than 5% and at cement replacement levels under 25% little difficulty is encountered in entraining air contents of 4 to 5%. Slight increase in the dosage of the admixture will usually adequately compensate for the air-detraining effects of such concretes. However, when 6 to 7% air contents are required, not only must the carbon content and cement replacement levels be below 5% and 25%, respectively, but also special care must be used to ensure uniformity. Slight changes in the consistency of the mix may cause large changes in air content. Consequently, good supervision and field testing of the concrete is required [4].

Other fine materials which tend to inhibit air entrainment include pigments, particularly carbon black. This is of concern to the ready-mixed operator supplying integrally coloured concrete for exterior exposure [5].

Calcium chloride which is used in cold weather concreting (in North America) can be used successfully in air-entrained concrete, but should be added separately and in solution form to the mix. Direct contact of admixtures containing high levels of calcium chloride with some types of air-entraining agents mixed in the same water phase may adversely affect both admixtures.

An understanding of the effect of variables on air entrainment is best

Table 5.1 Variability of concrete for dams produced containing air-entraining agents, water-reducing agents, or a combination of both

Dam	Type of admixture	Coefficient of variation (%) of compressive strength at 3 months
Okutadami Tagokura Hatanagi No. 1	AEA only	12–15
Kurobe No. 4 Hitotsuse Moromaki Arimine	AEA and WRA	8–10.5
Sakamoto	WRA only	5

shown by a consideration of the data collected during the construction of some dams in Japan [6]. The following relationship was found to exist where the concrete was air entrained either by an air-entraining agent or an air-entraining agent with a water-reducing admixture:

$$\Delta\sigma = -3\Delta s - 5.9\Delta a$$

where, $\Delta\sigma$ is the variation on compressive strength (MPa) at 91 days, Δs, the variation in slump (mm) and Δa, the variation in air content.

An increase in slump of 25 mm reduced compressive strength by 3 MPa. Thus, the control of air content is necessary to obtain reasonable standard deviation or coefficients of variation in compressive strength. In the study cited the compressive strengths at three months shown in Table 5.1 were obtained.

It has sometimes been argued that concrete is seldom damaged by frost action if low water–cement ratio mixes are used. However, the paramount effect of air entrainment in improving freeze–thaw resistance was recently clearly demonstrated in a study of the freeze–thaw resistance of both air-entrained and non-air-entrained superplasticized low water–cement ratio (< 0.35) concrete. The investigators concluded that the effect of providing an adequate air void system in concrete far overshadowed any improvement in freeze–thaw resistance as a result of lowering the water–cement ratio [7].

The method of placing air-entrained concrete can vary from simple

Fig. 5.3 Placing of a concrete for a carriageway using a slip-form paver.

Fig. 5.4 Finished carriageway.

manual placing between retaining side forms to more complex slip-form operations where the consecutive procedure of concrete placing, compaction, forming, finishing and insertion of joints is carried out by one or more machines. Figures 5.3, 5.4 and 5.5 show placing of a concrete highway and sidewalk by the slip-form technique.

Air-entrained concrete is also used in pumping. Concretes with air contents of 4–6% can be satisfactorily pumped, though some trouble may be experienced at the high end of the range since the elastic compression of air on each stroke reduces the efficiency of the pump. This is particularly likely to occur with long pipelines. Notwithstanding this limitation, it has been shown that air entrainment assists pumping in more cases than otherwise [8]. The air content of the placed concrete is not significantly affected by pumping, although small losses have been noted. Air entrainment is important especially if the concrete without air is deficient in fines and therefore inclined to be harsh.

Air entrainment plays a significant role in providing the required plasticity for lightweight aggregate mixes. The texture of some lightweight aggregates tend to make the concrete harsh, requiring an increased fine aggregate content. Consequently, the density of the resultant concrete is increased. Air entrainment enables the percentage of fine aggregate to be kept low, helps to prevent the coarse aggregate particles from floating in the mortar fraction, and facilitates pumping of such mixes.

A major application of air-entraining agents in the precast industry is in the production of lightweight cladding panels. High air contents are incorporated in the panels to provide proper coarse aggregate distribution and enhance durability.

The use of superplasticizers in air-entrained concrete has caused much debate. Two main problems are associated with superplasticized air-

Fig. 5.5 Placing of a concrete sidewalk using a slipform paver. (Photograph – Courtesy, Barber Greene, USA)

entrained concrete: (i) a decrease in air content by 1 to 3% when slump is increased from 75 mm to 220 mm after the addition of the superplasticizer to create flowing concrete and (ii) a change in the air void system to less desirable values. However, most investigators [9, 10, 11] have shown that, although the air void spacing factor required for adequate frost resistance is altered, the change did not necessarily affect the freeze–thaw durability of concrete.

The decrease in air content in flowing concrete is due to the release of entrapped air as a result of reduced paste viscosity. The reduction is usually offset by an increase in the dosage of the air-entraining agent [9, 11] to compensate for the lost air content. The addition of the air-entraining agent after the blending of the superplasticizer in the mix has been suggested as a suitable means of minimizing increases in the air void spacing [12].

Air-entraining admixtures confer significant benefits to concrete produced in hot weather countries such as Central and South America and the Middle East. The admixture is widely used to upgrade poor quality sands by compensating for deficient fine material. Consequently, bleeding and segregation are reduced, e.g. air entrainment eliminates the considerable separation of the mix (into an upper supernatant layer and lower aggregate layer) that occurs in cast *in situ* columns when sands deficient in particles less than 150 μm diameter are used in the mix. Figure 5.6 [13] illustrates this.

The variability in strength of air-entrained concrete is greater than that of plain concrete. It is, therefore, prudent to use the air-entraining agent in combination with a water-reducing admixture, or to use a single composite air-entraining and water-reducing admixture, since it leads to a lower standard deviation than a plain air-entraining admixture. Also, as the mean

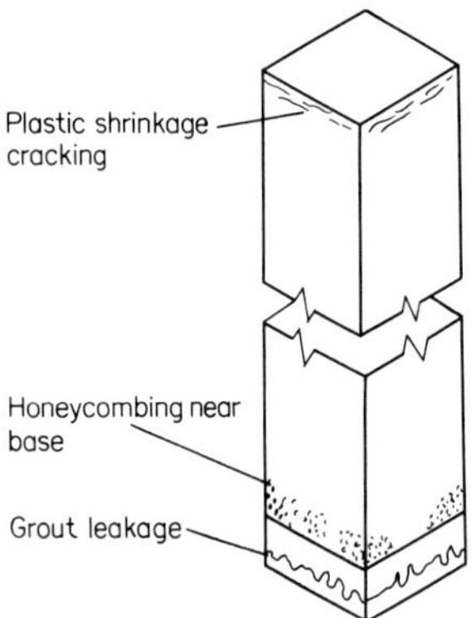

Fig. 5.6 Diagrammatic representation of problems arising from poor aggregate qualities. Air entrainment can improve the quality of concrete columns, particularly where the fine aggregate is of sea-dredged origin

Table 5.2 Air-entrained concrete has a higher standard deviation than plain concrete, but this can be minimized by the use of an air-entraining, water-reducing agent

Type of admixture in 270 kg m^{-3} OPC mix	Mean strength (N mm^{-2})	Standard deviation (N mm^{-2})
None (plain mix)	31.8	5.2
Neutralized wood resins	31.6	8.6
Air-entraining water-reducing agent	33.4	7.9

strength at an equivalent cement content is higher, the use of the admixture will result in a more consistent product. This is illustrated in Table 5.2, where the mean strengths and standard deviations of batches from one plant using plain concrete, air-entrained concrete with neutralized wood resins, and composite air-entraining and water-reducing agent are presented.

Because of the essential role that air-entrained concrete has played in reducing freeze–thaw susceptibility, its use in North America has been predominantly associated with this property of hardened concrete. Consequently, other favourable modifications have often been ignored. In Europe, Australia and Africa where freeze–thaw action seldom or rarely affects the concrete, the secondary modifications produced by air-entraining admixtures have been realized to their full potential, when used as a composite water-reducing and air-entraining admixture. The increased use of this type of admixture is due to the following reasons:

(i) Frequently specified maximum water–cement ratio of 0.55 with the required mix cohesion and minimum strengths are readily attained at reasonable cement contents.
(ii) The lower water–cement ratio and higher cohesion produced at the desired compacting factor values in the mixes, provide a more 'structured' concrete under conditions of no applied force. This is of particular importance in slip-formed concrete where reduced edge slump is required.
(iii) The higher mean strength obtained results in a concrete with a lower rejection rate than corresponding concretes containing conventional air-entraining admixtures only. Typical results of comparative mixes containing no air-entraining agent, a normal neutralized wood resin, and a composite water-reducing and air-entraining agent are shown in Table 5.3, where the benefits obtained are clearly illustrated.

In North America almost all air-entrained concrete contains a water-reducing admixture. There is some reluctance to use a composite type of admixture because of the concern that it reduces flexibility in use due to the variation in concrete materials and job site conditions. Therefore, the use of a composite water-reducing and air-entraining admixture for the purpose of improving product consistency finds little use in North America.

Table 5.3 Comparative properties of plain concrete and air-entrained concrete produced with neutralized wood resins and a water-reducing air-entraining agent

		Control	Mix 1	Mix 2
Admixture		None	Traditional neutralized wood resin	Water-reducing air-entraining agent
Mix design	Aggregate type	40–10 mm rounded gravel	40–10 mm rounded gravel	40–10 mm rounded gravel
	Sand	Zone 2	Zone 2	Zone 2
	A/c ratio	6.0	6.0	6.0
	% fines	33	33	33
	W/c ratio	0.59	0.56	0.54
Properties of plastic concrete	Slump (mm)	45 mm	40 mm	45 mm
	Compacting factor	0.89	0.91	0.93
	VeBe (s)	5.0	3.5	2.0
	Air content	1.0	3.9	4.1
Properties of hardened concrete	Average compressive strength			
	$N\ mm^{-2}$ (psi) 7 day	30.1 (4300)	28.0 (4000)	30.8 (4400)
	28 day	37.3 (5335)	32.5 (4650)	37.2 (5320)

5.3 Water-reducing and retarding admixtures

5.3.1 Ready-mixed concrete

In Canada and the United States the ready-mix concrete industry is by far the largest user of cement and accounts for well over half the total consumption. In the UK and most Western European countries 40–50% of the concrete output is also ready mixed [14]. Ready-mixed concrete is now widely used in many sophisticated ways including low slump concrete for highway paving, heavyweight concrete for thermonuclear plants, low heat concrete for prestress work, lightweight concrete for high rise buildings, and gap graded concrete for exposed aggregate architectural treatments.

In general, most of the ready-mixed concrete used in North America contains an admixture. Since freeze–thaw resistance plays a significant role in the durability of the concrete, the combined use of water-reducing and air-entraining agents is common. In Europe the use of air-entraining agents is less frequent. The ready-mixed concrete producer in both continents usually supplies concrete containing admixtures either on request from the client to provide a specific material or as a means of providing a specific type of concrete, e.g. watertight concrete or as a means of achieving the most economic use of mix ingredients.

Although the use of admixtures by the ready-mix sector of the industry is generally similar to that of site batched concrete, there are several unique elements and problems in control: in situations where concrete is specified by compressive strength and workability, normal and retarding water-reducing admixtures are widely used as a means of attaining the required properties at lower cement contents than mixes which contain no admixture; in ready-mixed concrete, reducing variation in concrete properties in both the plastic and hardened state from batch to batch is of considerable importance in minimizing rejection levels in field batches.

In this context, admixtures play a role in the maintainence of uniformity. This is illustrated in the following examples.

Table 5.4 presents a comparison of the results obtained from concrete batches produced on the same plant with and without admixtures. The table summarizes data collected over a six month period for two concretes of differing slump values (50 mm and 70 mm). It can be seen that the hydroxycarboxylic acid based normal water-reducing admixture produced no effect on the standard deviation for the 50 mm slump mixes, whilst an increase is noted for the higher workability mixes.

The effect produced by the incorporation of a lignosulphonate based water-reducing agent is shown in Table 5.5. The results were obtained from a series of mixes over an eight month period by a ready-mix plant used in the

Table 5.4 Changes in standard deviations of a ready-mix concrete plant using a hydroxy-carboxylic water-reducing agent

Mix	Slump (mm)	Admixture	No. of results	Average 28-day strength (N mm^{-2})	Standard deviation (N mm^{-2})
1	75	No	59	46.0	4.3
	75	Yes	61	52.0	5.8
2	50	No	386	44.0	5.0
	50	Yes	43	48.0	4.9

production of concrete piles. Since the standard deviation of this particular plant was 5.0 MPa for mixes produced without the use of admixtures, it is evident that the use of the admixture resulted in reduced variability.

These results indicate that in high workability mixes with cement contents in the median range, the admixture may cause an increase in the standard deviation. Thus in re-designing the mix to have a lower cement content in this class of concrete, adequate consideration should be given to this difference in standard deviation. Increased uniformity can be attained in this instance if the increase in standard deviation is compensated for by not utilizing the full potential cement reductions indicated by the mean 28-day strengths.

Figure 5.7 shows statistical data for strength tests from a ready-mixed concrete plant [15]. The coefficient of variation of 13.7% indicates an operation with a fair degree of control. However, these were random strength tests taken during the placing of foundation, sidewalks, driveways and miscellaneous construction typical of the small user of ready-mixed concrete. The results represent concrete mixes where there was wide variation in slump, sand gradation, moisture content, mixing time and where a high coefficient of variation (20%) is usually anticipated. The significant difference between this and the usual concrete delivered to the small consumer is that a water-reducing admixture was used throughout.

Another example of the effects produced by admixtures in ready-mix concrete is also shown in Fig. 5.8 in which data from another ready-mix plant is presented. The slope lines show that a change in slump from 75-175 mm

Table 5.5 7-day and 28-day strengths and standard deviations for concrete containing a lignosulphonate water-reducing agent produced on a ready-mix concrete plant

No. of mixes	Mean strength (N mm^{-2}) 7 day	28 day	Standard deviation (N mm^{-2}) 7 day	28 day
53	55.4	66.4	4.6	4.2

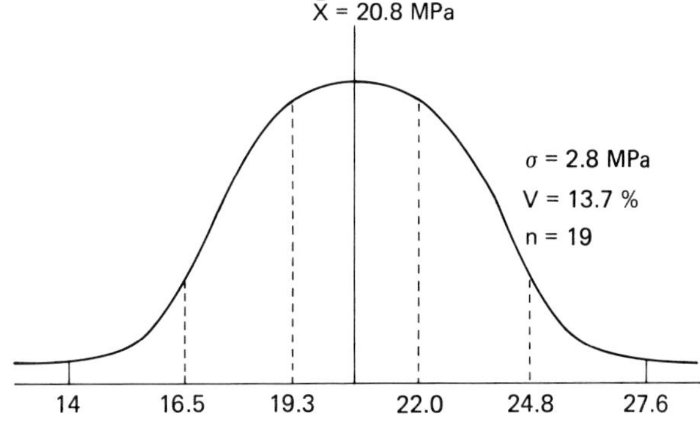

Fig. 5.7 Frequency distribution of control tests from a typical ready-mix concrete plant (Howard *et al.* [15]).

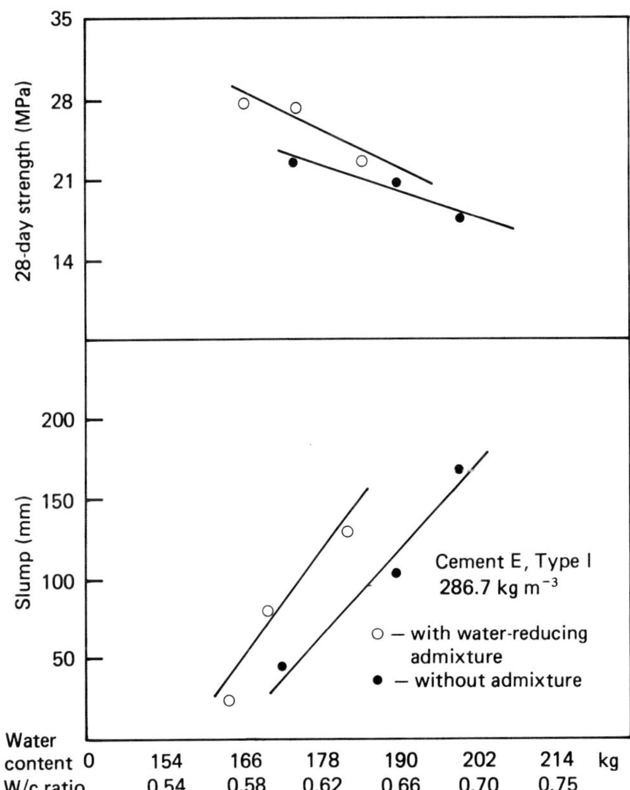

Fig. 5.8 Influence of water-reducing admixture on water content, slump, water–cement ratio (Howard *et al.* [15]).

without a water-reducing admixture required an increase in water–cement ratio of 0.08. With the admixture the same variation in slump required an increase in water–cement ratio of only 0.05, indicating that such concretes permitted variations in slump with less than the usual variation in water demand and water–cement ratio.

5.3.2 High strength concrete

High strength concrete is routinely produced in precast and prestressed applications through the use of low slump rich mixes which require prolonged vibration or shock methods for consolidation. However, site batched concrete utilizes less robust forms than those used in precast concrete and therefore the same compaction procedures cannot be used. Thus, plastic and workable concretes are necessary to avoid segregation or honeycombing[16].

Aggregate–cement bond and matrix strength play a significant role in determining the strength of high strength concrete. The high cement contents that are generally required for such mixes are often counter-productive. High shrinkage stresses produced cause loss of aggregate–cement bond or cracking of the cement paste due to the restraint induced by the aggregate particles. Matrix strength is primarily dependent on matrix porosity, which is governed by the water–cement ratio [17].

The increased plasticity and reduced water and cement contents required to achieve high strength can be attained using normal and retarding water-reducing or superplasticizing admixtures. In general, water-reducing retarders (based on hydroxycarboxylic acids) are more effective in such mixes than normal water reducers.

Depending on the type of admixtures used, two approaches are feasible:

Method A Using a normal or retarding water reducer, the water–cement ratio is reduced 6–10% at the same cement content and slump.

Method B Using a superplasticizer, a lower water–cement ratio is produced at a lower cement content with the same increased workability.

Table 5.6 shows the results obtained with a hydroxycarboxylic acid using Method A, while Table 5.7 shows the results obtained with superplasticizers using Method B.

5.3.3 Large pours and retarded concrete

The placing of large volumes of concrete in successive layers to form a monolithic structure is carried out in the construction of dams, bridge piers, bridge decks and caissons for jetties. Since the thermal properties of the concrete play a significant role in the differential volume change of mass

Table 5.6 High strength concrete produced by a hydroxycarboxylic acid water-reducing agent

Mix details		Control	Test mix 1	Test mix 2
Rapid hardening cement (kg)		500	500	500
Zone 2 sand (kg)		785	785	785
Crushed rock (10–5 mm) (kg)		1320	1320	1320
Hydroxycarboxylic acid water-reducing agent (ml/50 kg cement)		0	70	140
Water–cement ratio		0.46	0.42	0.41
Properties of concrete				
Slump (mm)		35	35	45
Air content (%)		0.8	0.8	0.7
Compressive strength at:	1day	19.3	21.0	20.7
	7 day	40.5	50.5	51.6
	28 day	49.3	59.5	62.8
Density (kg m^{-3})		2415	2432	2437

concrete, mixes for the construction of dams have mainly low cement contents [18].

Water-reducing retarders can contribute to the quality, economy and ease of placing of mass concrete. Projects such as dams which require substantial amounts of concrete are commonly located in remote areas of rough terrain where the cost of cement is high. The use of a water-reducing retarder permits water and cement reduction, improved workability and resistance to segregation of the mix during transportation. The lower cement content made possible may also decrease the heat generation and temperature rise in the structures (see later).

Unlike the lean (low cement content) concrete used in dams, the construction of bridge piers and bridge decks involve pours as large as 3000–4000 m^3 of high strength concrete. Consequently, the potential for development of damaging internal stresses in reinforcement or cracks in the structure is increased. However, special techniques developed to offset such stresses allow pours of large magnitude to be placed without adverse effects on the concrete [19, 20]. Among the techniques used in such pours is the use of water-reducing retarders which can make the following contributions [19]:

(i) It is recognized that success in reducing thermal stresses is dependent on minimizing the temperature differential between two points in the concrete; a maximum temperature differential of 20 °C has been suggested [19]. Therefore, any reduction in temperature rise will augment the measures taken to reduce this differential. Although water-reducing retarders themselves will not lower the total heat evolved, the lower cement content made possible by the use of admixtures decreases heat generation and temperature rise. In

Table 5.7 Cement reduction and change to type I with superplasticizers (La Fraugh [69])

Cement		Water–cement ratio	Admixture		Slump (mm)	VeBe (s)	Compressive strength (MPa)			Remarks
							18 hour		28 day	
Type	Content (kg m^{-3})		Type	Dosage			21 °C cure	60 °C cure	21 °C cure	
III	415	0.37	Zeecon	228 ml	57	3.8	24.24	40.34	75.50	Control
III	386	0.36	Mighty 150	1.2%	51	3.9	29.23	44.75	77.16	
III	386	0.33	Sikament	2.5%	38	5.8	29.10	49.71	82.46	
I	386	0.34	Mighty 150	1.2%	57	3.8	17.72	39.65	72.33	
I	386	0.31	Sikament	2.5%	114	2.1	22.62	41.99	81.02	

*Zeecon in millilitres per 50 kg cement.
Superplasticizers in percent of liquid by weight of cement.

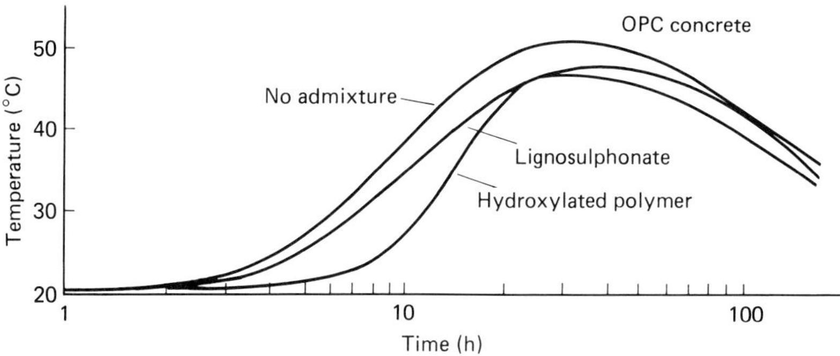

Fig. 5.9 Water-reducing admixtures can be used to reduce cement contents whilst maintaining the required compressive strength. In large pours reduced temperature rise is beneficial in minimizing the risk of cracking due to thermal expansion under restrained condition (Browne).

addition, admixtures delay the liberation of heat and reduce the peak temperature attained during hydration [21, 22].

Figure 5.9 gives results for the heat output for mixes containing 385 kg m^{-3} cement in a control concrete containing no admixture and 330 kg m^{-3} cement in a concrete containing an admixture [20]. The reduction in maximum temperature is a favourable effect which supplements other measures taken to minimize the temperature differential.

(ii) In bridge construction the use of retarding admixtures produces uniformity in the rate of setting and lessens the risk of deflection in partially hardened concrete that may occur in continously reinforced structures such as spandrel beams. The retarded concrete poured over the supports remains plastic until the final pours are placed at mid-span.

(iii) Continuous girder bridges are another type of structure where set control plays an important role. The weight of the concrete as it is successively placed on the deck deflects the girders causing the partially hardened concrete to crack.

Conventional concreting procedures consist of placing alternate panels so that a positive moment results (Fig. 5.10) [23]. The method is time intensive and costly to the contractor since finishing operations are intermittent and also involve periods of waiting for the concrete to achieve adequate strength. The use of retarded concrete provides a situation where deflection occurs when the concrete is still plastic and capable of deforming. No cracking occurs and the finishing operations can follow the pour sequence.

(iv) Structures with a high degree of reinforcement are susceptible to loss of bond between steel and concrete. This is particularly important in such situations where the vibration required to consolidate the concrete is

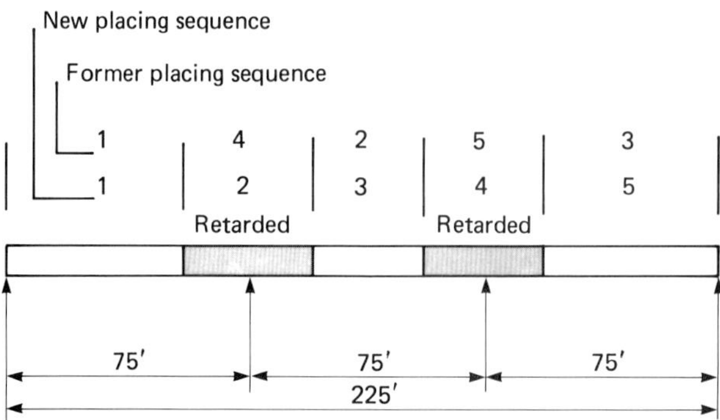

Fig. 5.10 Concreting sequence for continuous bridge deck. (Courtesy, R.J. Shutz [23])

transmitted along the reinforcement and could cause loss of bond in previously placed partially hardened concrete. The adjustment of admixture dosage as placement proceeds ensures that the concrete remains plastic throughout the entire pouring period.

(v) Retarding admixtures compensate for adverse ambient conditions particularly in hot weather concreting where they help to overcome the damaging accelerating effects of high temperature. The admixtures will lengthen the permissible period for vibration between batching and placing by an extension of the vibrational limit of the concrete (see later). Thus, as an operational procedure it is an advantage in such pours to use concrete retarders, since inevitable delays will not necessarily result in the loss of concrete that is being produced by rapid consecutive batching.

(vi) Retarding admixtures also help to eliminate cold joints and other discontinuities when concrete is placed in layers by enabling adjacent layers to be vibrated into each other. Since setting time governs the optimum time at which concrete can be re-vibrated, the slower the set, the more effectively can concrete be re-vibrated at later ages without loss of strength [23]. Figure 5.11 illustrates this effect.

(vii) If a pour is to be halted either by operational problems or by design, the last layer placed before the interruption can be further retarded by the use of larger dosages of the admixture. This can eliminate the necessity for construction joints. The heavily retarded concrete can be horizontal in the placing of layers or vertical when a sloping layer of concrete should be left.

Aspects pertaining to the use of admixtures are essentially the same for ready-mixed concrete as site-mixed concrete. However, when ready-mixed concrete is to be used for slip-forming or mass pour

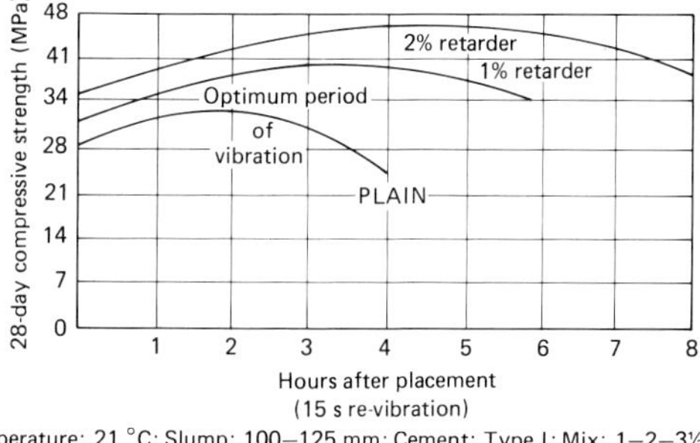

Fig. 5.11 The extension of revibration time of concrete using various dosages of a retarding admixture.
(Courtesy, R.J. Shutz [23])

operations, greater care and good judgement should be exercised in determining the dosage required to obtain the desired retardation time, because of the time involved in transportation from plant to site. In the case of large pours this requires continous planning, to provide delivery rates that meet the demands of the job site.

5.3.4 Marine structures

Concrete has been used in the construction of structures exposed to the sea for at least 2000 years [24]. However, it is in the decade of the 1970s, with the discovery of offshore gas and oil, that major advances in the utilization of concrete in a marine environment have been made [25].

Concrete exposed to a marine environment will be subjected to wave action which imposes dynamic load, shock, impact stresses, erosion by abrasion, freezing and thawing, wetting and drying and the chemical reaction of chlorides and sulphates. The durability of concrete under these conditions is correlated with low permeability and high strength. A number of recommendations pertaining to the requirements of concrete to be used in marine structures have been made [24, 25]. These include specification of cement contents of 400–500 kg m^{-3}, a water–cement ratio of less than 0.45 and an air content of 4–7% depending on the maximum aggregate size. Such parameters usually involve mix modifications which include an admixture.

Conventional water-reducing admixtures used in marine construction are normally of the set retarding type. The use of air-entraining agents in conjunction with water-reducers in concrete located at the splash zone area confers improved resistance of freeze–thaw cycling and sulphate attack. Typical basic mixes containing a water-reducing admixture for use in a marine environment for gravity platform construction are shown in Table 5.8.

Table 5.8 Characteristics of concrete used in the marine concrete structure shown in Fig. 5.13.

	Control	Mix 1	Mix 2
Mix details			
Ordinary Portland cement (kg m^{-3})	425	425	425
Graded PFA (kg m^{-3})	105	105	105
Sand zone 2 (kg m^{-3})	611	611	611
20/10 mm coarse aggregate	1053	1053	1053
Lignosulphonate WRA (ml/50 kg)	0	140	140
Hydroxycarboxylic acid WRA* (ml/50 kg)	0	140	400
Properties of concrete			
Slump (mm)	105	100	110
Proctor needle penetration resistance (h to 0.5 N mm^{-2})	8.5	12.5	27.0
Air content (%)	1.4	1.8	1.6
Compressive strength (N mm^{-2}) at 7 day	39.0	53.4	58.7
28 day	52.7	60.2	64.8
Density (kg m^{-3})	2410	2420	2460

*Level varied according to the slip-forming requirements of retardation.

Figures 5.12 and 5.13 show large structures for marine environments where water-reducing admixtures were used throughout the concrete together with retarding and air-entraining admixtures in part of the structures.

The construction of offshore structures is the most complex and demanding of all concreting operations. Working conditions are difficult

Fig. 5.12 This large marina involved the placing of large caissons; water-reducing admixtures of the lignosulphonate type were utilized in their manufacture.

Fig. 5.13 This gravity oil production platform contained admixtures in all 200 000 m^3 of concrete and utilized the benefits of flowing concrete for the precast segments, retarded concrete for the slipformed walls and a normal water-reducing agent to obtain maximum compressive strength and impermeability.

and delays and alterations in procedures occur more frequently. A variety of mixes incorporating two or even three admixtures are used. Notwithstanding these contraints, it is possible that with proper dispensing equipment and a good degree of control of the operation, a high level consistency in concrete properties can be obtained. This is illustrated by the histogram (Fig. 5.14) for the production of concrete for an oil production gravity structure.

In recent years superplasticizers have found increasing use in the construction of offshore structures. The significant effects on water reduction (>20%) and workability (slump of 200 mm) produced in concrete have enabled the ready placing of low water–cement ratio (< 0.45) mixes in heavily reinforced areas that are characteristic of these structures. The water reduction afforded by the use of superplasticizers helps to achieve the desired low permeability and high strength that are so important to ensure

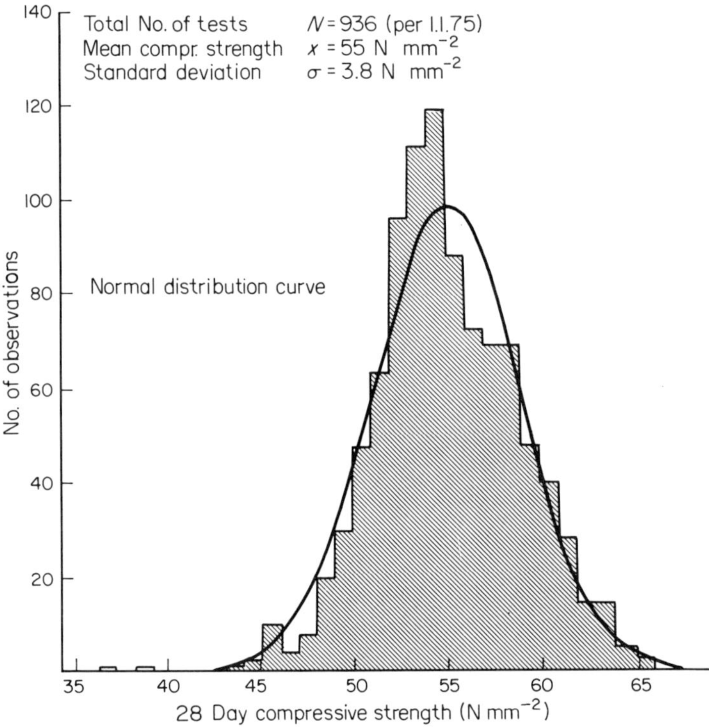

Fig. 5.14 The distribution of concrete cube tests for an oil production platform containing various admixtures.

durability and longevity in service under marine conditions. Figure 1.54 shows the favourable effect low water–cement ratio has on the permeability of concrete.

In the area of tidal concreting for sea defence work, e.g. between tides, the concrete is required to develop strength rapidly even under winter conditions. Until its exclusion from such structures, calcium chloride has been the major admixture capable of providing the early strength development required for this particular operation. However, a recent study [26] indicates that superplasticizers, when used in conjunction with silica fume, may offer a viable alternative to calcium chloride in this application. It was shown that the combined accelerating effects of silica fume and the superplasticizer produce higher strengths at low temperatures (5 °C) than that obtained from chloride containing concrete for admixture dosages of 1.5 and 0.6% w/w of cement of calcium chloride and superplasticizer, respectively (Fig. 5.15) [26]. In addition, superplasticized silica fume concrete improves other hardened properties of concrete such as density,

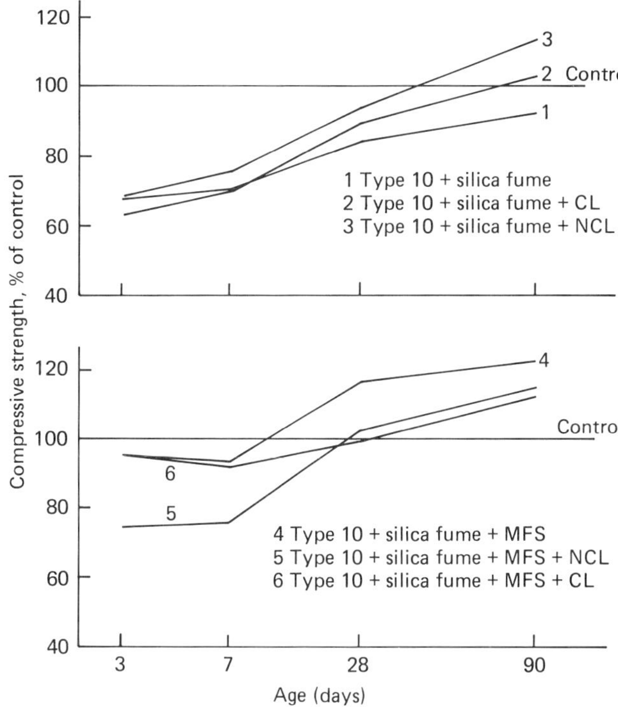

Fig. 5.15 Relative compressive strengths obtained for silica fume/Portland cement mixes containing superplasticizers (MFS – 3%), chloride (Cl – 0.6%) and non-chloride (NCl – 0.6%) admixtures at 5 °C.

abrasion, resistance to and freeze–thaw sulphate attack. This is useful for concrete exposed to sea-water.

5.3.5 Slipforming

While pavements are slipformed in the horizontal plane, the normal concept of slipform is in the vertical plane. Vertical slipforming is usually chosen over traditional concreting procedures for the following reasons:

(i) The method produces a high quality monolithic structure with a minimum number of joints.
(ii) It is fast and enables other phases of the job to be scheduled and completed more rapidly and efficiently.

For slipforming to be successful, continuous tightly scheduled material deliveries must be developed and then maintained to ensure uninterrupted concrete placement. Concrete of low to high strength and from low to high workability has been slipformed successfully using retarding water-reducers, superplasticizers and air-entraining agents under both summer and winter conditions. The technique has also been used for lightweight concrete.

In this system, vertical forms slowly slip up the semi-hardened concrete on jacks riding on steel rods extending from the concrete lift below. Once the form is filled and the slipforming operation is under way the form is kept filled, the rate of filling being timed to match the predicted rate of rise. The lifting rate used is such that a stable structure is left behind, since the primary feature of the slipform process is the ability of the concrete to support itself at an early age. Typical exposure times for the concrete after placing could range from 1.5 to 6 hours. The requirements for setting time of the concrete are dictated by the time and distance between batching and placing and the time taken for concrete to emerge from the slipform. In warm weather, retarding water-reducing admixtures, cooled mix water or fly ash are used to delay set so that sufficient time elapses to allow the transport and slipping of concrete. Conversely, when slipform operations are conducted under winter conditions, the desired slipping times may be attained by the use of Type III cement (RHPC), heating of the concrete to 22 °C in the forms, use of protective enclosures, and high dosages of an accelerating admixture. Air-entraining admixtures are also used in slipforming, not only to enhance durability characteristics but also to minimize the staining caused by excess bleed water running down the sides of the structure [27]. Table 5.9 shows mixes where both fly ash and admixtures were used to achieve the desired slipping times in the construction of the perimeter wall of a nuclear reactor building.

Mix proportions and set-controlling methods are usually determined from preliminary trial mixes conducted at the job site using actual construction materials to provide a guide for the slipping and set times required during the slipforming process. The required data are usually obtained by preparing penetration resistance charts (using the Proctor penetration resistance test, ASTM-C-403-20) which shows set-time variation with different admixture dosages (Fig. 5.16).

Variation of admixture dosage during the course of the slipforming process is determined by changes in temperature conditions. Each dosage change corresponds to the amount required to compensate for the effect of a change in temperature of 8.2 °C averaged between the concrete mix temperature and the air temperature [28]. The changes made should be incorporated in a gradual manner to avoid drastic changes in concrete setting time.

The Proctor penetration test is widely accepted for both initial laboratory trial and on site testing to determine correlations between setting

Table 5.9 Monthly statistical analysis and concrete mix design summary sheet

Area: Reactor building
Class: 3.5–3.7

Period ending: April 1976

				3.5		3.6		3.7	
	Mix designation			3.5		3.6		3.7	
	Period in use		From:	April 13, 1976		April 15, 1976		April 16, 1976	
			To:	April 28, 1976		April 28, 1976		April 27, 1976	
Design requirements:	Strength	MPa		34.5 (39.7)		34.5 (39.7)		34.5 (39.7)	
	W/c			0.395		0.394		0.397	
	Slump	± 12 mm		100		100		100	
	Air	± 1%		4.5		4.5		4.5	
	Sand	% of total agg. by vol.		35		35		35	
Mix design proportions:	Cement	kg m^{-3}		368		380		380	
	Fly ash	kg m^{-3}		83.1		83.1		83.1	
	Water	kg		178		182		184	
	Sand			575.5		575.5		575.5	
	Stone	mm 0.19		1074		1074		1074	
		mm 0.12		—		—		—	
	AEA	MBVR	ml	280		280		315	
	WRA	Pozz	ml	680 (100 × R)		440 (100 × R)		215 (100 × R)	
Statistical analysis:				Month	To date	Month	To date	Month	To date
	No. of samples – n			19	19	15	15	31	31
	Average strength – $\bar{x}$(MPa)			43.6	43.6	40.7	40.7	38.8	38.8
	Standard deviation – σMPa			3.38	3.38	2.62	2.6	2.3	2.3
	Coefficient of variation								
	Overall V			7.8	7.8	6.4	6.4	6.0	6.0
	Within v			2.2	2.2	2.9	2.9	4.4	4.4

Courtesy, New Brunswick Power Commission, Canada.

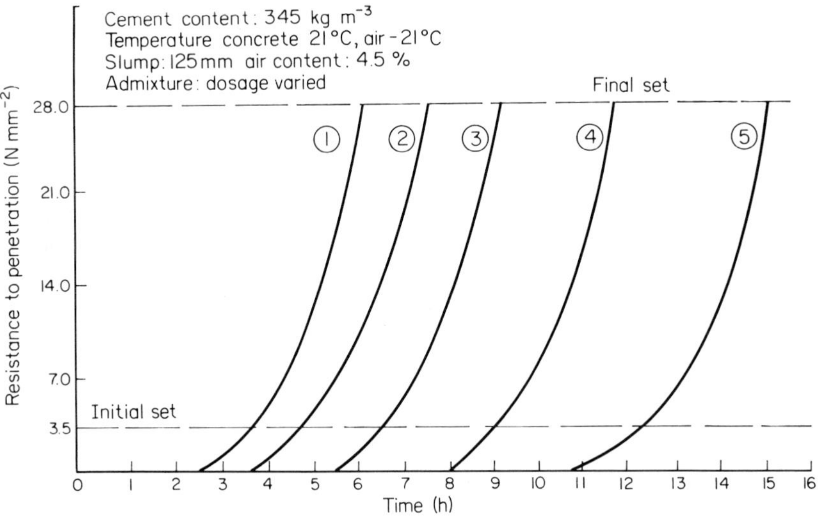

Fig. 5.16 Proctor needle penetration resistance for mortar sieved from concretes containing varying retarder levels.

characteristics of the concrete and the rate of slipping of forms. Penetration resistance values in the range of 0.7–3.0 MPa have been found optimal for successful slipforming. When using the Proctor penetration test as a means of estimating slipping times, care must be taken to ensure that storage conditions for the mortar containers are strictly comparable with those of the concrete in the slipforms. The effects of ambient temperature and high winds, heat generation, etc., should be taken into account [28].

Vertical slipforming is used in the construction of cooling towers, tapered chimneys, service cores for buildings, bridge piers, shaft linings and, more recently, in reactors for nuclear power stations. Typical structures slipformed from mixes containing hydroxycarboxylic acid, carbohydrate polymer type water-reducing and air-entraining admixtures are shown in Figs 5.17 and 5.18 respectively.

5.3.6 High workability mixes

In the construction industry the term 'high workability' in concrete is applied to concrete with slumps above 100 mm that are readily consolidated by vibration. Conventional lignosulphonate, hydroxycarboxylic and carbohydrate polymer based water reducers usually produce slumps of up to 150 mm with little adverse effect, provided the fines content of the mix is increased to maintain cohesion. Such mixes find use in areas of congested reinforcement and thin sections with poor access, such as piling and diaphragm walling. These admixtures offer an alternative to the traditional methods which required increase in both cement and sand contents of the mix. Figure 5.19 shows a typical application for high workability concrete where a lignosulphonate water reducer was used.

More recently the use of admixture systems which use high dosages of two or more admixtures have been promoted for mixes requiring high workability. The system usually consists of the use of a combination of admixtures based on the concept of (a) complimentary and (b) compensatory action between two materials with similar functions; for

Fig. 5.17 A slip-formed tower using concrete retarded with a hydroxycarboxylic acid based water-reducing agent.

Fig. 5.18 Slipforming of the perimeter wall of a nuclear reactor building. (Photograph – courtesy, New Brunswick Power Commission, Canada).

example, in the production of high early strength the increased rate of strength gain achieved by water reduction through use of a normal setting water reducer can be augmented by accelerated hydration produced by the use of an accelerator. Additionally, the adverse side effect (excessive set retardation) that results from the use of high dosages of the water reducer can be offset by the set-acceleration effect produced by an accelerating admixture. Likewise, combinations of a water reducer and retarder or water reducer and a low dosage of superplasticizer may be used to effect economy in the production of flowing concrete (see later).

5.3.7 Pumping

The advent of small bore pumping units has made possible rapid placing of concrete, providing significant savings in manpower and cost. Concrete can be transported from a delivery point to a point of placing in a single operation with minium labour, regardless of elevation or site obstructions, using pumping techniques.

Fig. 5.19 The congested reinforcement in this application necessitated the use of high workability concrete. A lignosulphonate based water-reducing agent was used.

Normal concrete mixes resist being moved under pressure due to segregation of the cement paste that occurs when pressure is applied. Basically a pumpable concrete will have a suitable aggregate–void system and a cement paste consistency that will flow adequately through the void channels. Both these properties of the mix are required to meet the primary demands of pumpability which are:

(a) The pump pressure must be transmitted to the solids.
(b) A continous annular grout film must be formed adjacent to the pipe wall to ensure sliding of the concrete core.

Such requirements can often be met by specially designed mixes for individual jobs, almost always involving a compromise in sand/aggregate ratio and water content. However, even with good mix design the incidence of blockage is high. Variation in concreting materials (aggregate shape, gradation and moisture content) account for the major pipeline blockages. Furthermore, today many specifiers will not allow alterations to mix proportions merely to accommodate pumping because of adverse side effects on the properties of hardened concrete. A more viable approach to the design of mixes to be pumped is offered by the use of admixtures.

Chemical admixtures broaden the envelope of aggregate gradations which may be used in the mix, enable concrete to be placed under a wider range of job conditions, and enhance the physical properties while making the mix more pumpable.

Three broad classes of pumpable concrete usually used are:

(i) Low cement content mixes (210 kg m^{-3}).
(ii) Medium cement content mixes (200–300 kg m^{-3}).
(iii) High cement content mixes (> 300 kg m^{-3}).

Mixes in both low and high cement content classes are more prone to problems than the medium range. In low cement content mixes poor cohesion results in segregation and in high cement content mixes thixotropy causes pipeline friction. Admixture will modify the flow characteristics of the paste, helping to achieve and maintain optimum flow characteristics.

For low cement content mixes, the admixture imparts water retentivity to the cement paste under forces tending to separate the mix water. Special admixtures are available for assisting pumping of low cement content mixes, but these materials are not included in this book. More information on these admixtures can be obtained by consulting the literature [29].

Concretes in the median range, although having satisfactory paste flow properties, often run into problems due to a lack of supply of consistent quality aggregates. Common problems are decreased cohesion of the cement pastes for mixes in the lower cement content range and increased friction to flow in mixes in the higher cement content range. In both instances the use of either normal or retarding water reducers in combination with an air-entraining agent will alleviate these problems. Air entrainment increases the cohesion of the cement paste, while the retarding water reducer enables the release of water to reduce the friction that develops in a thixotropic paste.

Mixes of high cement content tend to have thixotropic pastes. Consequently, flow through the aggregate–void channels is inhibited and the mobility of the peripheral grout layer decreases. Admixtures used in this class of concrete are of the dispersing agent type which induce lubrication by an increase in the free water content of the mix. Commonly used materials are calcium lignosulphonates and sodium salts of hydroxycarboxylic acid.

Pumped concrete must not only meet specified job performance criteria (e.g. strength, freeze–thaw resistance) but should also remain stable under a variety of job conditions, particularly in hot and cold weather. It is, therefore, common to find that concrete to be pumped will often contain two or more types of admixtures.

Pumping of lightweight concrete is another area where admixtures play a significant role in improving pumping characteristics. Such concretes are inherently more susceptible to segregation and absorption of water under pressure than normal concretes. The use of admixtures (air-entraining agent, superplasticizer or thickener) imparts increased viscosity and plasticity to the mix resulting in improved pumpability.

5.3.8 Underwater concrete

The most common method of placing concrete under water is by tremie. A tremie (Fig. 5.20) consists of a vertical pipeline, topped by a hopper, which is long enough to reach the lowest point to be concreted from a working platform above the water [30, 31]. The tremie method is widely used for casting mass concrete in caissons, cofferdam seals and bridge piers.

Concrete mixes used for this purpose are usually cement rich (> 400 kg m^{-3}) with slumps in the range of 150–225 mm. They possess good cohesion and flowability. The fine aggregate proportion is usually higher than for normal concretes (about 45% total aggregate content) and the coarse aggregate size restricted to a maximum of 40 mm. Placing is continuous and no compaction by vibration is permitted. The lower end of the tremie pipe is always kept submerged in the fresh concrete to maintain a seal and to force the concrete to flow into position by pressure.

The quality and strength of tremie concrete is greatly dependent on proper mix design and placement. Concrete poured in this manner usually has poor edges and much laitence. Flow problems can be encountered, extending the placement time and causing gravel pockets due to the need to lift the pipe to facilitate concrete flow. Cohesiveness and flow properties are greatly improved by using admixtures [32, 33].

A water-reducing retarder and air-entraining agent are generally recommended as a combination that enhances uniformity of the placed concrete. More recently, the use of superplasticizers, which provide greater flowability and cohesion at significantly lower water content, has superseded the use of conventional admixtures in this field. Previous work [30, 32, 33, 34] has indicated that the use of admixtures in this application realizes the following advantages:

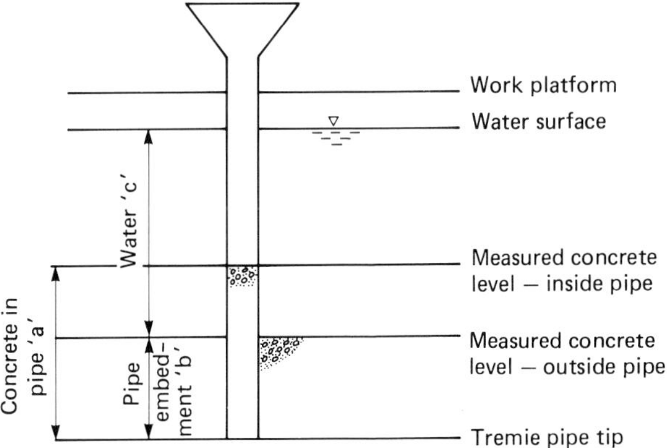

Fig. 5.20 Outline of a tremie pipe (Gerwick *et al.* [31]).

(i) Increased flowability at a given water content allows greater horizontal movement (9–11 m) compared to concrete without admixtures. Flatter slopes are produced enabling wider spacing of the tremie pipes (Fig. 5.21). The rate of placing is increased due to increased hydraulic pressure; concrete, therefore, moves further and faster. Increased pipe lengths are possible, consequently, substantial savings are achieved through the reduction of the number of tremies used.

(ii) Higher water reduction produces cohesive mixes less susceptible to segregation, laitence and wash-out of cement paste. Superplasticizers permit the use of wider pipes for faster placing and allow cement reductions so that less heat is generated.

(iii) Retardation of the rate of stiffening allows for longer placing times and wider spacing of tremie pipes.

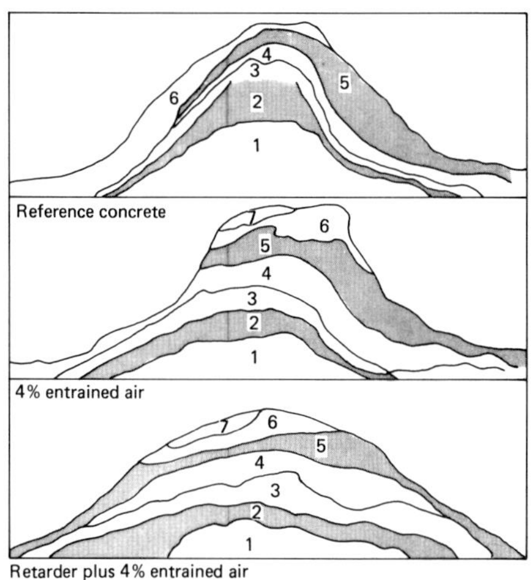

Fig. 5.21 Tremie flow patterns using concrete with and without admixtures [34]. Numbers indicate sequence of batches. Thick end sections are laitence and poor quality concrete. Note reduction in thick end sections for pour containing retarding and air-entraining admixture.

5.3.9 Piling

Cast-in-place piling has similar mix design requirement as pumping mixes, e.g. high workability with adequate cohesion to minimize segregation and bleeding during placing. The maximum coarse aggregate size is usually small (< 40 mm) and a high sand content is used. Although they are not subject to weathering exposure, such as may occur with precast piles, cast-in-place piles may be subject to aggressive water in the soil. Consequently, the water–cement ratio must be kept to a minimum [13].

The higher water demand of the mix (high fines content, small coarse aggregate) coupled with workability and durability requirements makes the use of an admixture essential. Conventional water-reducing admixtures and superplasticizers can readily provide lower water–cement ratios at the desired high workability.

5.3.10 Concrete subject to water under hydrostatic pressure

Concrete which is to be subjected to water under a hydrostatic pressure is often specified as being 'watertight' or 'waterproofed' concrete. As explained earlier, hydrophobic type waterproofing admixtures are ineffective in preventing the ingress of water under these conditions. By far the best method of producing concrete suitable for these applications is to utilize a good quality conventional water-reducing agent or a superplasticizer. Water reductions ranging from 10–20% are afforded by the use of these admixtures resulting in water–cement ratios significantly (< 10%) lower than that obtained in corresponding mixes without admixtures; therefore, a considerable reduction in concrete permeability is obtained. This is illustrated in Fig. 1.54 of Chapter 1. Water-reducing admixtures can result in a double improvement in the concrete when:

(i) The water-reducing admixture is used to produce half the potential water reduction attainable (say 5%) and the slump is increased to a higher value (say 125 mm) than the mix with no admixture.
(ii) When high dosages of normal water-reducing and accelerating admixtures are used, to produce high water reduction and high workability.
(iii) When high dosages of superplasticizer are used to effect low water–cement ratio and high workability.

The significantly lowered water–cement ratio and the efficient compaction that is achieved by the use of these admixtures is the best means of producing good quality impermeable concrete.

5.3.11 Tilt-up construction

The technique of horizontal on site precasting of wall panels (often up to heights of 12.2 m) is commonly referred to as tilt-up construction [35]. Ready-mixed concrete used for such purposes is of two types, structural and architectural. Architectural concrete, in addition to possessing structural properties, may require a specific mix parameter (e.g. exposed aggregate finish) to produce a variety of surface textures.

Admixtures are generally used to provide the required flexural strength (2.1 MPa) and design tilt-up strengths (compressive) with permissible minimum cement content, and/or to modify setting characteristics of the mix. Normal set-retarding and superplasticizing agents are used to obtain the desired water reduction for high strength concrete (> MPa). Accelerators are usually used in cool weather or when the concrete is exposed to high ambient temperatures and wind conditions. The latter use of accelerators is particularly helpful in preventing crusting and plastic shrinkage. Air-entraining agents are used to provide durability under freezing–thawing and wetting–drying conditions.

5.4 Superplasticizing admixtures

Modifications afforded by the use of superplasticizers are essentially marked extensions in performance of the basic function of conventional admixtures. The chemicals used do not significantly affect the surface tension of water in the mix. Consequently, they can be used at higher dosages without the attendant adverse side effects of high air entrainment and set retardation encountered with normal plasticizing and water-reducing admixtures. The two main applications for superplasticizers are as follows:

(a) Flowing concrete: where slumps in excess of 180 mm are used and the admixture is added to the mix with no alteration in water–cement ratio.
(b) Water reduced high strength concrete: where low to moderate slumps (25–75 mm) are used at reduced water–cement ratios to attain high early and ultimate compressive strengths.

5.4.1 Flowing concrete

Such concretes usually possess high consistencies with slump values in excess of 180 mm and flow table spread greater than 50 cm. The high workability is usually achieved by the addition of the superplasticizer to a 50–75 mm slump as seen in Fig. 5.22.

Due to the fluid-like character of flowing concrete mixes, there is a tendency for increased bleeding and segregation when normal mixes with slumps in the range of 75–100 mm are raised to values in excess of 180 mm. Therefore, some alteration to mix design (Chapter 1, p. 56) is required to maintain adequate cohesion of mix. Aspects to be considered in the design of such mixes are:

(i) Cement type and content.
(ii) Fines content of the mix.
(iii) Aggregate properties.
(iv) Maximum placing slump.

(a)

(b)

Fig. 5.22 (a) 50 mm slump prior to the addition of a superplasticizer; (b) Collapsed slump of 100 mm generally used in flowing concrete.
(Slides – Courtesy, S.K.W. Trostberg, W. Germany)

(v) Dosage of the admixture (as determined by admixture type, cement type concrete temperature and initial slump).
(vi) Sequence of addition.

Cement types IV and V are reported to require lower admixture dosage than types I and III to produce a given slump. The fineness of the cement may influence both the degree of slump increase and the strength levels attained. Finely ground cements require higher water contents or increased dosages to reach the desired high workability. The optimum cement contents which provide flowing concrete have been found to be in the range of 270–375 kg m^{-3} [36, 37].

Coarse aggregate characteristics such as shape and texture should be considered. Mixes containing crushed or angular aggregates will require a higher proportion of fines. A decrease in maximum aggregate size will usually promote flow character.

The use of concretes with slumps $>$ 220 mm or flow table spread $>$ 60 cm is not recommended since these mixes are prone to bleeding and segregation, particularly under vibration or when conveyor belts are used to transport the concrete.

Due to the varying solids content and, hence, the effective ingredient concentration in the various commercially available superplasticizers, particular attention should be paid to the manufacturer's recommended dosage for flowing concrete.

Factors which affect the dosage rate are concrete temperature, initial slump (i.e. slump before the addition of the superplasticizer), cement type and content, the presence of other conventional admixtures in the mix prior to the addition of the superplasticizer and the sequence of addition to the mix.

High concrete temperature, finely ground cement, high cement content (> 415 kg m^{-3}) and low initial slumps will require higher admixture dosages than the manufacturer's standard recommended dosage.

The presence of an air-entraining agent, retarder or water reducer in the mix will produce a higher than anticipated slump increase. Consequently, bleeding and segregation may occur due to the cumulative dispersing action of the two admixtures. This behaviour is more prone in mixes with lower cement and fines contents. Control of all these variables ensures that consecutive loads are similar in their placing and handling characteristics.

Flowing concrete has revolutionized concrete pumping techniques. High workability and concommitant cohesion achieved through the use of superplasticizers enables concrete to be placed farther, at faster rates and lower pumpline pressures. Figure 5.23 shows the dramatic effects produced by incorporating superplasticizers in pumped concrete.

Another area of application where the advantages resulting from the use of flowing concrete is obvious, is in the construction of floor slabs, roof decks and concrete bay areas. The distribution of concrete over large areas and

Fig. 5.23 The use of a superplasticizer in pumped concrete. (Photograph – Courtesy, S.K.W. Trostberg W. Germany)

areas with congested reinforcement or constricted form configuration is both time consuming and labour intensive. However, the use of a highly mobile mix such as flowing concrete will reduce the time and labour required for placement and consolidation.

The typical advantage of the mix mobility attained in these mixes is illustrated in Fig. 5.24 which shows the construction of a base limit for a water treatment plant. The closely placed plastic pipes precluded the use of normal vibratory compaction due to poor vibrator access and possible damage to the pipes. The highly workable concrete used flowed around the pipes allowing for both speed and efficiency in placing. Table 5.10 shows the mix used and the results obtained.

Although the modifications obtained and the resulting properties of the concrete are essentially the same in all types of concrete, ready-mixed concrete involves a delay between the mixing and placing of the concrete and on-site addition of the admixture. Consequently, the time-dependent properties of the plastic concrete and the manner in which the admixture is incorporated into the mix requires good control.

For ready-mixed concrete which is to be produced in a highly workable form it is necessary to deliver to the site concrete having a slump of 75 ± 10 mm and add the correct plasticizer dosage using a suitable dispenser. Although manual dosing is often used, the type of dispenser shown in Fig. 5.25, which fits into the chute truck, is recommended. In some situations where the concrete is centrally plant mixed, the superplasticizer dosage to be added for each load is calculated at the plant and carried in a container on the mixer truck, (Fig. 5.26). The required dosage is then fed under gravity through a tube into the mixing drum [38]. The calculation is generally based on charts prepared (prior to the pour) for the superplasticizer dosages required to produce non-segregating flowing concrete at various slumps. This enables the required dosage adjustments to be readily made when difficulties in providing the requisite slump on site are encountered. The method usually requires sufficient control of consistency and relies on minimum periods of delay and, therefore, requires tight scheduling.

After the addition of the superplasticizer, flow table or slump cone measurements should be taken and recorded. Although the flow table

Table 5.10 Mix details and concrete characteristics for flowing concrete in Fig. 5.24

Ordinary Portland cement (kg m^{-3})		=	350
10 mm crushed limestone (kg m^{-3})		=	1009
Zone 2 marine sand (kg m^{-3})		=	925
W/c ratio		=	0.52
Slump (before admixture addition)		=	50 mm
Compressive strength (N mm^{-2}) at:	7 day	=	23.0
	28 day	=	28.0

(a)

(b)

(c)

Fig. 5.24 (a), (b), (c) Construction of a base limit for a water treatment plant.

Fig. 5.25 A superplasticizer dispenser which fits into the chute of the ready-mixed concrete vehicle.

method is accepted by many European countries, North American workers [39] suggest that it is extremely sensitive to operator error and prefer the use of the slump cone. They contend that the initial value to spread after taking off the slump cone is as sensitive a measure of mobility as the final spread. More recently, an apparatus called the free orifice rheometer [40], which measures flowing properties of the concrete, has been introduced to the market.

The use of these admixtures in this segment of the industry has been slow to gather momentum in North America and is usually restricted to a small number of more sophisticated ready-mix plants which supply mixes in situations where other materials have failed or in areas which require special concreting techniques. High admixture cost, their potential contribution to slump loss and their unique effect on the properties of fresh and hardened concrete have been the main constraints to their more widespread adoption. By contrast ready-mixed concrete operators in Europe and Japan, who experience the constraints of tight scheduling, have found flowing concrete a boon in meeting their objectives.

The main benefit of flowing concrete is to the contractor who is willing to absorb the initial cost for the special concrete, recouping these costs in his own operations and savings in placing costs. There are also advantages to the

Fig. 5.26 Calculated dosage of the superplasticizer carried in a container on the truck and added by the gravity fed method.
(Slide – Courtesy, Flowmix, Canada. Cement Lafarge)

ready-mix concrete industry since the quicker placing allows faster truck turn-round time and shorter standing time on site. Indeed, the ready-mix concrete plant must adjust its shipment to meet the increased demands of productivity on the building site in order to avoid stoppages due to delays in concrete delivery; otherwise the savings in labour costs resulting from increased placing rates may be dissipated. It is particularly significant in jobs requiring placing of concrete in foundations or footings where it is often necessary for the truck to make several manoeuvres to complete placement. The greater radius of flow (about 5 m) of this type of concrete from the point of placing considerably reduces the number of manoeuvres required. Wherever possible the chute should reach as far towards the middle of the placing area as possible. Extension tubes can be added to increase the discharge range accordingly. The use of such tubes enables concrete to be placed further, as well as into inaccessible parts of the job. This is illustrated in Fig. 5.27 where concrete slabs are being placed by the use of traditional 50 mm concrete by hand raking and finishing, in comparison to a slab poured with flowing concrete.

The unique characteristics of flowing concrete in which high workability

(a)

(b)

Fig. 5.27 The placing of slabs (a) by the use of flowing concrete and (b) by the traditional method.

Fig. 5.28 Cores taken from the slabs of Fig. 5.27, after 3 years. (A) Flowing concrete. (B) 50 mm slump concrete.

and high cohesion co-exist, enables the coarse aggregate to be evenly distributed throughout the height of the placed concrete. This is clearly illustrated in the cores shown in Fig. 5.28.

The cores which were obtained three years later from the pour illustrated in Fig. 5.27 gave the results presented in Table 5.11. It is interesting to note that the increased strengths obtained from the cores taken from flowing concrete slabs can be correlated with the lack of voids (Fig. 5.28, core A).

The range of applications for which flowing concrete is used includes diaphragm walling, slab on grade construction and casting heavily reinforced structural elements.

The areas where the designer can use highly workable (flowing) ready-mixed concrete are as follows:

(i) In difficult pours with congested reinforcement and poor vibrator access.
(ii) In large pours where construction joints can be reduced thus speeding up construction.
(iii) In complex shuttering (forms).

Table 5.11 Characteristics of the three-year-old cores shown in Fig. 5.28

	Slab from flowing concrete	50 mm slump concrete
Visual examination		
Distribution of materials	Good	Good
Compaction of concrete	Good	Poor–fair
Voids: large	None	Few
medium		
small	Negligible	Considerable
Test results		
Corrected cylinder strength ($N\,mm^{-2}$)	32.8	26.3
Estimated cube strength ($N\,mm^{-3}$)	40.8	33.0
Density ($kg\,m^{-3}$)	2330	2305

(iv) In smaller structural elements where there is economy in design because of higher strengths attainable.

One of the significant limitations of the use of ready-mixed flowing concrete is the rapid reduction in the high workability that is initially achieved. For ready-mixed concrete using truck agitators greater slump loss was found when there was (a) a long delay in the time between addition of superplasticizer and sampling, (b) a long delay in the time between mixing and addition of superplasticizer, and (c) when concrete temperatures were significantly higher than ambient temperatures.

A number of methods have been used to minimize slump loss problems in ready-mixed concrete. Of these the most successful methods include the following [12]:

(i) The use of a hydroxycarboxylic acid based retarding admixture in conjuction with the superplasticizer.
(ii) The use of a retarding superplasticizer .
(iii) Incremental addition of more admixture to restore the original slump [41].

Method (iii) has been successfully used on jobs where the specification excludes the addition of any water on the job site [42]. More recently, products have become available which by their chemical nature allow considerable extension of the period of workability.

Despite wide publication of the adverse effects that re-tempering causes, job site addition of extra water to compensate for slump loss is still a common practice, particularly in hot weather conditions. Under these conditions the use of both convential and superplasticizing admixtures helps to minimize the amount of water required for re-tempering so that the loss of strength is minimized [43, 44]. This is shown in Table 5.12.

Table 5.12 Effect of water-reducing admixtures on re-tempering (Previte [43])

Cement B*		Time (min.)		
		10	120	126
Reference	Slump, mm (in)	92 (3⅝)	35 (1½)	98 (3⅞)
0.18% modified lignosulphonate	W/c	0.56	0.56	0.62
	Slump, mm (in)	92 (3⅝)	41 (1⅝)	114 (4½)
0.75% hydroxylated carboxylic acid	W/c	0.52	0.52	058
	Slump, mm (in)	95 (3¾)	38 (1½)	95 (3¾)
1% s/s sulphonated melamine formaldehyde	W/c	0.49	0.49	0.56
	Slump, mm (in)	92 (3⅝)	35 (1⅜)	92 (3⅝)

*Temperature: 21 °C (70 °F).
Water-reducing admixtures allow a significant reduction in total water after re-tempering.

Flowing concrete can be utilized in the precast industry to aid production of complex shaped precast concrete units. Often it is the only solution to a problem posed by such situations, e.g. in the construction of toroid-shaped dome units for the Ninian Platform in Scotland, high early strength had to be rationalized with proper compaction in heavily reinforced components. The solution in this instance was the use of a high strength flowing concrete produced with the help of a superplasticizer. This not only enabled proper consolidation to be achieved but also provided the high early strength required to maintain a fast turnover of moulds. Figure 5.29 shows construction of the heavily reinforced dome units.

Fig. 5.29 Construction of the heavily reinforced toroid shaped dome structures for the Ninian Platform in Scotland.
(Slide – Courtesy, S.K.W. Trostberg)

Fig. 5.30 The intricate detailing of these precast concrete panels necessitated the use of flowing concrete.

Flowing concrete also offers significant advantages to architectural concrete where texture is required. Complex shapes are produced by casting flowing concrete on textured form liners to obtain the desired effect. Figure 5.30 shows a building incorporating cladding panels with ribbed and grit blasted external surfaces. This particular example has intricate reverse tapers which were produced using a 25–40 mm superplasticized concrete that also permitted early stripping of the unit.

Many concrete producers use a combination of superplasticizer and a conventional water-reducing, set-modifying or air-entraining admixture to achieve the desired performance. The superplasticizer provides the major portion of the required water reduction and the conventional admixture is added to achieve one or more of the following objectives (a) further water reduction, (b) admixture economy, (c) the desired air content, (d) increased workability, and (e) extension of set and workability. Commercial formulations used for this purpose are usually based on sodium lignosulphonates, hydroxycarboxylic acids or processed carbohydrates. Such combinations, besides reducing the dosage of the superplasticizer required, are reported to improve mix workability by reducing the high mix cohesiveness and consequently, the finishing effort required in some mixes [12].

As previously described, slump loss constitutes one of the chief constraints to wider acceptance of superplasticizers; therefore, a number of major producers of admixtures have sponsored active research to improve the workability retention characteristics of their superplasticizers. Some recent developments [45, 46] have shown promise, and among these are materials based on lignosulphonates.

Materials composed of selected lignosulphonate fractions of high molecular weight, low in carbohydrate content, have proved successful and are efficient dispersants with acceptable retardation characteristics. At reduced dosages in comparison with the more widely known products such as melamine and naphthalene sulphonates, the former materials give a comparable performance with respect to workability retention. Some degree of air entrainment results from the use of these materials and strength values are marginally lower than those obtained with other

Fig. 5.31 Very workable but cohesive concrete can be produced from most aggregates and within the normal limits of workability control by the use of air-entraining water-reducing agents based on lignosulphonates.

superplasticizers. More work is needed prior to the wide commercial use of such materials.

Lignosulphonates with some air-entraining capacity have been used at higher than normal dosages to produce highly workable and cohesive mixes. By modifying the mix design to maintain or increase the cement content or by using an accelerator, little or no deleterious effect on strength is obtained and yet cohesive concrete can be obtained from mixes having initial slumps falling in a range much wider than that required for superplasticizers. Figure 5.31 shows a close-up view of this type of concrete while Table 5.13 gives the physical properties.

Table 5. 13 Mix details and concrete properties of highly workable concrete produced using a lignosulphonate-based air-entraining, water-reducing agent

	Basic mix	Modified mix for flowing concrete
Mix composition		
Cement OPC (kg m^{-3})	280	300
Coarse aggregate (kg m^{-3})	1300	1300
Sand 50/50 zones 1 and 4 (kg m^{-3})	700	600
Air-entraining water-reducing admixture (ml/50 kg cement)	0	300
Concrete properties		
Slump (mm)	65	70 (before addition)
Flow table spread (cm)	—	56 (after addition)
Air content (%)	0.9	4.6
Compressive strength (N mm^{-2}) at: 7 day	18.4	15.9
28 day	32.8	30.1
Density (kg m^{-3})	2450	2405

5.4.2 High range water-reduced concrete

The greatly improved workability that results when a superplasticizer is added to concretes made at normal water–cement ratios, makes it possible to produce concrete having normal workability (75–90 mm slump) but with significantly reduced water–cement ratios. Superplasticizers are claimed to provide water reductions in the region of 20–30% [36, 47, 48] depending on the type of cement used, aggregate properties and concrete temperature. In contrast, conventional water reducers provide reductions of only 10–15%. The water content is progressively reduced as the dosage of the superplasticizer is increased at a given workability. This trend is observed for mixes with cement contents in the range of 300–600 kg m^{-3} [47, 48]

The water reducing and strength increasing effects produced by superplasticizers varies with the type and brand of cement used. With some cements superplasticizers contribute to greater water reduction but not significantly higher strengths. The reasons for this relative performance have not been fully determined. However, some factors that may account for difference in behaviour are the C_3A and SO_3 contents and the Blaine fineness of the cement [49, 50, 51].

The proportion of coarse to fine aggregate, the fineness modulus of the fine aggregate and the maximum size of the coarse aggregate affect properties such as percentage water reduction, early and ultimate strengths, placing and finishing characteristics of such concretes.

Water reduced cement rich mixes containing superplasticizers produce mortar rich cohesive concretes which respond slowly to applied vibration, often exerting a strong frictional drag on finishing tools. Vibrator effectiveness and reduced cohesion can be produced by increasing the coarse aggregate fraction and/or using a more coarsely graded (F.M. 2.8 - 3.1) fine aggregate [52].

The dosage of the superplasticizer required will depend on the admixture type, the amount of water reduction required, the desired placing slump, the cement type and content, the concrete temperature and the presence of other conventional admixtures in the mix prior to the addition of the superplasticizer. The actual values used for initial and placing slump are usually governed by parameters such as desired early strength levels, durability criteria, the degree of vibrator compaction to be used and the finishing characteristics required for the application.

In general, normal slumps at lower admixture dosage may be produced with retarding superplasticizers, cements with low C_3A contents ($< 5\%$) [53, 54], mid-range cement contents [36, 54], coarsely ground cements, and the presence of the other conventional admixtures in the mix. Low initial slump (< 50 mm) will usually require higher admixture dosage to provide conventional slumps.

Calcium lignosulphonate based water reducers produce complementary effects when used with superplasticizers. However, they should be added to the mix separately, the former with the gauging water and the latter at the end of the mix cycle. The combined use of the two admixtures results in the following modifications in comparison with similar mixes which contain only the superplasticizers: (a) Higher water reduction and extended initial set times for mixes with slump < 100 mm. For mixes with slumps > 200 mm, depending on the dosage of the normal water reducer used, significant set extension and some strength reduction. (b) The presence of the lignosulphonate admixture makes the mix more susceptible to segregation on re-dosage of the superplasticizer [55]. The cumulative dispersing action is higher for naphthalene formaldehyde than the melamine formaldehyde superplasticizers.

When used with superplasticizers, hydroxycarboxylic acid (HC) based retarding admixtures confer benefits to concrete particularly under high ambient temperature conditions [50, 54] and in steam curing in precast operations [50, 53]. Depending on the dosage of the HC admixture, the concrete temperature and placing slumps, an extension of 3–4 h. in the setting time should be expected.

Both neutralized vinsol resin and sulphonated hydrocarbon type air-entraining agents are compatible with all types of superplasticizers. The four main problems encountered with air-entrained flowing concrete have also been identified for water-reduced superplasticized concrete. However, with the exception of reduced air content in low slump mixes, the magnitude of the other problems is much reduced in comparison with flowing concrete. The reduced air content in low slump mixes is caused by the lack of a minimum paste visosity required for the air-entraining mechanism. Some producers have overcome this problem by adding the air-entraining agent after the superplasticizers so that the required viscosity is provided [52].

Most of the work on superplasticized Portland cement concrete containing fly ash or blast furnace slag has shown that such mixes require 10% less admixture than reference Portland cement concrete to attain the same workability. Therefore, a given dosage may produce higher water reduction. The reason for the reduced admixture requirement has not been determined. It is probably due to the lowering (dilution) of the C_3A content of Portland cement that results when fly ash or slag replaces part of the cement in a concrete mix.

Common objectives sought in the water reduced high strength application include:

(a) Lowering of curing cost.
(b) Reduced cement content while maintaining same workability and strength levels.
(c) The use of an alternative cement type, e.g. from Type III to Type I.
(d) The production of more workable concrete with high early and ultimate strengths.

These are discussed below:

(a) Lowering of curing cost

Superplasticizers afford substantial reduction in the curing temperature and the length of the curing cycle normally used for accelerated curing of concrete products, e.g. when cement contents are not reduced significant water reduction is effected, resulting in high early strengths under elevated curing temperatures. This advantage has permitted some producers to reduce the maximum curing temperature by 50–60 °C and to reduce the curing period by 3–4 h. while maintaining strength levels comparable to concrete cured in a conventional manner [52, 53].

If the superplasticizer used is of the non-retarding type which is added to the mix at the normal recommended dosage, no modification of the set time results and the normal curing cycle may be used. However, a modification to this curing cycle may be necessary for higher superplasticizer dosages or when a retarding superplasticizer is used, or when the superplasticizer is used in combination with a normal water-reducing retarder. In these situations the final setting time may be lengthened, and preset period may need to be increased accordingly. The length of the total curing cycle may not necessarily be increased [50].

(b) Cement reduction

At their normal dosage rates most commercial superplasticizers can produce slumps comparable to that of a plain mix while reducing mix water content up to 20–25%. The lower water–cement ratio may then be used to increase existing compressive strengths or to reduce cement content.

The use of superplasticizers may enable the precast producer to use lower cement content without reduction in mix workability and rate of strength development. The actual amount of cement reduction achieved will depend on the cement type used and the mix proportion used in the concrete. Previous work [56] indicates that even with low cement content (306 kg m^{-3}) a normal dose of superplasticizer can accelerate 3- and 28-day strengths by 90% and 55%, respectively, over levels attained with a plain mix. Cement reduction in the range of 11–20% have been achieved in mixes with a cement content of 415 kg m^{-3}, while maintaining desired strength levels [53]. An example of the use of a superplasticizer admixture to achieve cement reduction was presented in Table 5.7.

(c) Use of alternative cement

Most often precast producers use finely ground cement such as ASTN Type III to obtain desired production cycles. The high water reduction and concomittant rapid strength gain afforded by the use of superplasticizers will enable the precast producer to change to the more economical Type I cement (Table 5.7).

(d) Production of more workable concrete with high early and ultimate strengths

The concrete slump used in any particular prestressed and precast operation is usually determined by balancing the opposing requirements that it be readily placeable but also that it achieve maximum strengths. The use of superplasticizers is a particularly effective means of reducing the effort required to place and consolidate low to moderate slump concretes. When consolidated with high acceleration vibrators the superplasticizer will cause even extremely low slump concrete to flow readily.

Superplasticizers are often used in water reduction application to increase 12–18 h. strengths. The substantial reduction in the time required to achieve stripping strength allows for early destressing of wire reinforcement and increased mould turnover for precast items.

High strengths realized by the use of superplasticizers have been used advantageously in the production of high strength precast columns in high rise buildings and precast girders. Here, the admixture permits the maximum attainable strength by effecting high water reduction, increasing mix workability, and contributing to the more efficient use of available cement.

A typical application where the three desired objectives of high water reduction, high workability and high early and ultimate strengths are simultaneously required is in nuclear concrete. The concrete for the construction of prestressed concrete pressure vessels for nuclear power plants is required to provide high early and ultimate strengths and high workability at low water–cement ratios. Additionally, it must exhibit the lowest possible thermal expansion. Consequently, special concretes incorporating high water demand aggregate such as limestone are required. The high water reduction and substantial plasticizing action afforded by superplasticizers is often the most convenient manner of achieving these objectives.

5.5 Accelerating admixtures

The set time of concrete can be significantly shortened and early strength increased by the use of accelerating admixtures. A wide range of soluble and inorganic salts have been used.

The benefits of an increase in strength gain are the following:

(a) To compensate for the slower hydration of cement at low temperatures.
(b) To permit earlier removal of forms.
(c) To reduce the required curing and protection period.
(d) To advance the time when the structure may be used.

Shortening of the setting time affords the advantages of:

(i) a reduction of hydraulic pressure on forms, and
(ii) the more effective plugging of leaks against hydraulic pressure [57].

The above objectives can be obtained by other means, namely: the use of cements with high Blaine fineness (Type III), the use of higher cement content, the use of heated materials, higher curing temperatures, or a combination of these. However, it is often found that the desired results can be more conveniently and economically achieved by the use of an accelerating admixture.

5.5.1 Calcium chloride

The most widely used and cheapest accelerator is calcium chloride, either by itself or as a main ingredient of some commercially available accelerators. Calcium chloride is available in two solid forms: Type I regular flake, containing 77% $CaCl_2$ and a concentrated pellet or granular form, containing 94% $CaCl_2$. It is also available as a liquid containing 32–40% solids.

Calcium chloride dissolves readily in the gauging water and can, therefore, be added to a concrete batch in dry form. However, the potential for problems arising from errors in batching and lack of adequate mixing make the previously prepared solution the preferred form. For this purpose a standard solution is recommended.

The beneficial effects of calcium chloride in early strength increase and the reduction of the setting time are relatively more significant when the admixtures is used at curing temperatures below 20 °C. The rate of heat evolution is increased, thus offsetting the usual slowing down of hydration reactions that occurs when low curing temperatures are used.

In North America, where there is only limited restriction on the use of calcium chloride in reinforced concrete, the use of admixtures to offset cold weather effects has depended almost entirely on chloride based admixtures. However, the growing concern for the effects of these admixtures on the corrosion of metals embedded in concrete has led to increasing restriction on their use. Thus, there is now a need for non-chloride based accelerating admixtures.

Calcium chloride has been excluded from reinforced or prestressed concrete by recommendations contained in a British Standard Code of Practice (CP 110) since 1977. Similar stipulations have been in force in France and Germany for the last ten years.

Although calcium chloride remains the most effective accelerator its use in concrete often results in adverse effects on some mechanical and durability properties (Chapter 4). ACI Committee 201 suggests limits for chloride ions for concrete exposed to different service exposures [58]. The user should employ these limits in conjunction with other known limitations of the use of calcium chloride to determine its compatibility and dosage in any given application.

The banning of the use of calcium chloride in reinforced concrete has stimulated the development of a number of alternative materials that emulate its accelerating properties without having the same potential for corrosion.

5.5.2 Chloride-free accelerators

Most of the commercially available chloride-free accelerators in Europe and North America are based on calcium formate. The admixture is usually used in powder form at a dosage ranging from 1–2% by weight of cement. Due to its very low solubility in water (16 g/100 g H_2O) the calcium formate is not commercially available in liquid form.

Other chemicals which have been used to a limited extent in the field include sodium nitrite, sodium thiosulphate, sodium sulphate, sodium tetraformate and calcium nitrite. The latter has now been commercialized in North America and Japan.

Calcium nitrite appears to be the most promising non-chloride accelerator to date. The admixture is marketed in liquid form as a 20% solution, and its capacity to be readily added to the mix using the usual admixture dispensers has proved fruitful for its acceptance by the ready-mix concrete industry. The strength development effects produced in concretes containing this admixture are reported to be comparable to those obtained with calcium chloride [59]. Table 5.14 shows a comparison of the results obtained for concrete containing calcium chloride, calcium nitrite, and no admixture. In addition to improvements in both compressive and tensile strengths, calcium nitrite is a significant corrosion inhibiting admixture. It is finding increasing acceptance in bridge deck repair work for this reason.

Table 5.14 Acceleration of strength development with calcium nitrite (Rosenberg *et al.* [58])

Cement brand	Admixture % by weight of cement	3-day compressive strength, (MPa)	(Psi)	Setting time Initial h : min	Final h : min
A	None	10.5	1525	8 : 45	12 : 21
A	Calcium nitrite 1%	10.8	1568	6 : 00	10 : 20
A	Calcium chloride 1%	14.8	2151	4 : 20	7 : 30
A	Calcium nitrite 2%	13.2	1924	3 : 05	6 : 55
A	Calcium chloride 2%	18.10	2624	2 : 10	4 : 55
B	None	10.11	1467	8 : 38	. . .
B	Calcium nitrite 1%	10.86	1576	5 : 24	9 : 05
B	Calcium chloride 1%	15.30	2220	3 : 16	5 : 00
B	Calcium nitrite 2%	14.30	2075	3 : 12	5 : 42
B	Calcium chloride 2%	17.66	2562	2 : 15	3 : 40

Water–cement ratio was 0.56 to 0.57; slump was 4.0 ± 0.5 in. (100 ± 12 mm); air content was 1.95 ± 0.25%.

5.5.3 Accelerators for use in Portland cement/fly or Portland/ cement blast furnace slag concrete

The use of fly ash and ground blast furnace slag in ready-mixed concrete in countries like Canada, the USA and Australia has significantly increased in the last decade. Consequently, a number of admixture manufacturers are marketing admixtures which are claimed to have the ability to catalyse the slag or fly ash/Portland cement reaction. Benefits claimed by the manufacturers include upgrading the rate of strength development in the early stage (especially the 1–3 day period) so that the concrete is brought in line with the performance expected from Portland cement concrete, the use of normal form stripping procedures, and enabling higher levels of fly ash or slag substitution. The chemical compositions of such admixtures are closely guarded trade secrets. There is, therefore, a paucity of published information with respect to the type of chemicals used and the mechanism of the reaction involved in the acceleration of the reaction.

Fig. 5.32 Efflorescence adversely affects the aesthetics of precast concrete items. The problem is particularly noticeable in dark pigmented items.

5.6 Damp-proofing admixtures

Although the usefulness of integral damp-proofing admixtures in imparting resistance to water penetration to concrete is minimized in watertight concrete, the significant role these admixtures play in maintaining the aesthetic qualities of architectural concrete cannot be disputed. The enduring effect of the admixture was discussed earlier (Table 3.8), where it was shown that the incorporation of such materials provides more aesthetically acceptable concrete products, which inhibit disfiguring algal growth. These admixtures also reduce the passage of rain or ground water through the concrete structure by capillary action and will often reduce unsightly efflorescence. The latter application is of particular significance in concrete products where colour is involved. The admixture in this instance curbs the transport of lime to the surface by reducing the ingress and passage of moisture. Figure 5.32 illustrates the ungainly deposit of white salts that occur due to efflorescence.

5.7 Concreting in hot weather and hot arid climates

In climates where hot weather prevails for most of the year, the contrast between minimum night and maximum day temperatures and humidities is often great. Most countries that experience such climates lie in the equatorial belt, where drastic geological conditions and the available raw materials are of poor quality. Under these conditions concrete is adversely affected unless it is designed and placed with special hot weather precautions.

The actual temperatures experienced (> 40 °C) go beyond the scope of many current recommended practices. For example the ACI Guide for hot weather concreting (ACI 305R-77) does not meet the needs of, and is not practical for, concrete mixing operations in such climates. Furthermore, the ASTM C-494 specification for chemical admixtures for concrete does not provide adequate information for the selection of admixtures for hot weater concreting since the evaluation is based on only a minimum mixing time at room temperature. The assumption that an admixture that retards at 21 °C will do so at 33 °C is not valid. Common retarders at 21 °C often act as accelerators at 33 °C [60]. Simulation of field conditions in the laboratory is usually not possible because of a multitude of varying ambient conditions. The best place for evaluation is, therefore, the actual job site, using the materials intended for the concrete.

Common problems encountered in hot weather concreting (particularly in countries of the Middle East, Central and South America) include the following:

(a) Decreased strengths due to increased water demand, resulting from temperatures and the use of local raw materials.
(b) The rapid decline in workability due to high slump loss, posing re-tempering problems.
(c) Difficulties in the handling and finishing of concrete as a result of accelerated setting due to increased rate of hydration and moisture loss from the mix.
(d) The increased tendency for plastic shrinkage caused by rapid evaporation due to high mix temperature and low relative humidity.
(e) Increased tendency for cracking due to drying, shrinkage and differential thermal movements.
(f) Decreased durability due to the presence of deleterious salts in aggregates.
(g) Difficulty in maintaining desired air contents as a result of high mix temperatures.
(h) The increased tendency for bleeding and settlement resulting from poor aggregate gradation.

Typical concrete parameters specified to offset the damaging effects of hot weather conditions include: minimum cement contents, minimum 7- and 28-day strengths, low water–cement ratio, maximum slump, and proper curing.

The contractor, in attempting to provide a concrete meeting the desired strength requirements and stipulated durability criteria, is often compelled to accept the time-consuming placing of concrete due to poor workability of mixes of low water–cement ratios; other factors include the use an uneconomical mix design to achieve high strength margins and the tolerance of a high percentage of rejected units having severe blemishes, due to poor compaction of dry mixes. For the contractor, therefore, adherence to specifications with conventional concreting practices results in high cost to the project.

Studies have shown that customary methods of increasing cement content to ensure strengths and durability characteristics, cause some of the deficiencies encountered in both precast and *in situ* concreting in hot weather countries. Further, accumulated data also show that the prudent use of admixtures often provides cost savings in both concrete materials and labour, enabling the production of a better quality concrete [44, 61–63]. Some of the mix modifications sought to offset the adverse effects of the use of poor concrete materials and the limiting environmental conditions are as follows:

(a) High water-reduction and plasticizing effects for concretes where low water–cement ratios and minimum cement contents are specified.
(b) Increased cohesiveness in harsh mixes and mixes prone to bleeding.

(c) Improved watertight character of the concrete to prevent the ingress of water containing high concentrations of sulphate and chloride ions.
(d) An accelerated rate of strength gain in mixes using Type V (sulphate resisting) cements.
(e) A retarded rate of initial set and fall-off of workability and possible reinstating of initial workability.
(f) A reduced rate of heat release in the structure.

These requirements which are often antithetical can in most instances be supplied by the use of a chemical admixture. This aspect is discussed as follows:

5.7.1 High range water reduction

The high water demand that results from the use of warm concrete materials and high ambient temperature in hot weather concreting, make the achieving of desired water–cement ratios in high quality mixes (high cement content and low water–cement ratios) difficult. In such situations the use of a water-reducing retarder or a retarding superplasticizer will help in attaining the required workability at the specified water–cement ratio. Figure 5.33 shows the use of stabbits for breakwater construction purposes, where a retarding superplasticizer proved useful. Parameters such as early strength, good finish, workability and reasonable water–cement ratio (0.52) were all required in a mix which used TypeV cement and beach sand.

Fig. 5.33 Stabbits for use in breakwater application constructed with the help of superplasticizers under hot arid ambient temperature conditions.
(Photograph – Courtesy, N.P. Mailvaganam)

5.7.2 Workability loss

Two major causes of slump loss have been identified when concrete is produced under hot arid conditions. These are water loss due to the high absorption of the dry aggregates and evaporation of water from the mix. Values of 5–6% and 12% have been recorded for the former and latter, respectively [63].

The water loss which is reported to take place early has been suggested as the major cause of the decline in workability in hot climates. Since the workability loss is quite rapid it is argued that cement hydration makes little contribution to it [63]. Consequently, under such conditions the use of retarders that delay the hydration reaction may prove ineffective. Indeed, retarders based on hydroxycarboxylic acid (HC) which are potent set retarders and efficient dispersing agents have been shown to produce greater slump loss than control concretes. Lignosulphonate based materials, which have limited set retarding properties, in comparison show improved workability retention. The reasons attributed to the difference in observed behaviour are that HC type admixtures, by their strong dispersing action, free mix water which increases the degree of mix water evaporation. Lignosulphonates on the other hand tend to entrain slight amounts of air and are not powerful dispersing agents. Thus, they may tend to retain moisture and not increase evaporation from the concrete. However, because of their milder set retarding characteristics, they are not as effective as HC type admixtures at maintaining low slump values under hot weather conditions [64]. Figure 5.34 shows the effect of the above-mentioned admixtures on the rate of slump loss at 50 °C.

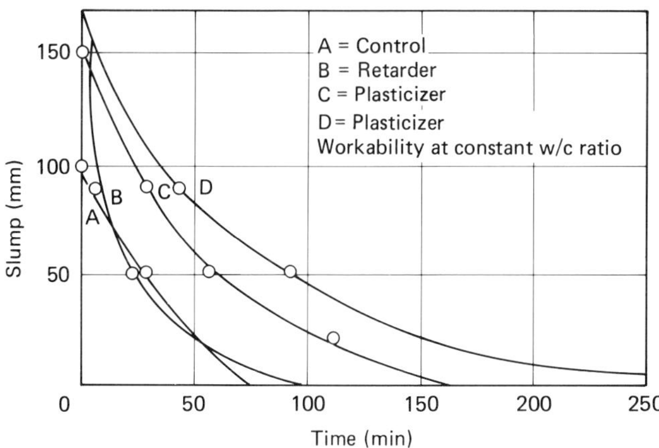

Fig. 5.34 Workability loss at 50 °C (McCarthy [63]).

5.7.3 Reinstating of initial workability

As previously mentioned, workability loss is a serious limiting factor in hot weather concreting. At temperatures exceeding 22 °C the mix is susceptible to rapid slump loss. Consequently, there are frequent cases where on-site re-tempering of the mix with additional water to provide the desired workability is carried out. The serious loss of strength that results from this practice warrants concern.

Although good concreting practice dictates the strong disapproval of on-site re-tempering of mixes, the practice continues in many sites which have poor control. Therefore, any action which offsets the potential for serious strength reduction deserves consideration. Two controversial methods have been suggested [41, 43, 44, 64]. These are:

(a) The use of retarders to enable late additions of water and remixing.
(b) The addition of a second dosage of a superplasticizer to restore the original slump.

Method (a) is sometimes carried out by ready-mix companies in hot climates when delays are encountered. However, the addition of further amounts of water is done under controlled conditions when the truck is brought back to the plant. The retarder delays the initial hydration reactions and, therefore, subsequent addition of water and remixing can be done without adverse effect on strength. Table 5.15 shows the effect of late addition of water and remixing a mix where a retarder was used, while Table 5.12 shows the use of both conventional and superplasticizing admixtures to minimize the amount of water required for re-tempering so that loss of strength was minimized.

For method (b) previous work [41, 65] has indicated that superplasticizers can be used effectively to reinstate the original slump without loss of strength. It is not certain as to how many subsequent additions can be made

Table 5.15 Effect of late addition of water and remixing of a mix containing retarder at 50 °C (McCurrich [64])

Admixture (hydroxycarboxylic)	3-day cube strength (MPa)
Control: Cube cast immediately and sealed	
None	21
Concrete remixed for intervals of 3½ h: water added to make good weight loss due to evaporation	
None	6
0.21 litres/50 kg OPC	20

before it becomes detrimental to the concrete. Laboratory investigations [41] and field experience [42, 50] indicate that, depending on the type of superplasticizer used, incremental dosage may result in either mild acceleration or retardation. The performance of each superplasticizer admixture in this respect is thought to be governed by an inherent response of the admixture to a given type of cement. The magnitude of the effect produced by re-dosage has been found to be closely related to the water–cement ratio [41]. Mixes with water–cement ratio less than 0.55 appear to respond favourably to re-dosage and attain initial high slumps with adequate cohesion. At slumps of 0.65 there appears to be a more drastic and cumulative effect whereby re-dosing at the same initial level produced severe bleeding and segregation. The effectiveness of a superplasticizer to increase slump appears to decrease as the age of the concrete increases, and this effect should therefore be considered when establishing the maximum period during which re-dosage can be carried out.

5.8 Cold weather concreting

In order to withstand damage that results under cold weather conditions concrete should attain a compressive strength of 5 MPa before exposure to freezing temperatures and 20.7 MPa before being subjected to freeze–thaw cycles [66]. Low temperature has a retarding effect on both the setting time and the strength development of concrete. Additionally, depending on the consistency of the mix, a reduced rate of hydration results in less water intake by the cement particles, promoting bleeding and segregation. Therefore acceleration of set and strength gain are desirable in concrete placed at low ambient temperatures.

Suitable precautions used to offset the limiting effects of cold weather include the following [67]:

(a) The heating of concrete materials.
(b) Protection of the placed concrete by insulating covers.
(c) The increase of cement content.
(d) The use of finely ground cements like Type III.
(e) The use of accelerating or superplasticizing high range water reducing admixtures.

Admixtures used under cold weather conditions are expected to fulfil two roles: to increase the hydration rate to offset the slowing down of the reaction that usually occurs as the curing temperature goes below 10 °C, and to reduce the amount of freezable water present in the concrete by affording good water reduction thereby minimizing frost damage. These are discussed below:

5.8.1 Acceleration of the rate of hydration

The effects produced on the hydration reaction by cold weather and accelerating admixtures are quite noticeable when heat of hydration curves are examined (Fig. 5.35). It can be noted that low temperature not only lowers the peak temperature but also significantly extends (shift of graph to the right) the course of the reaction.

Accelerators, on the other hand, increase both the peak temperature and markedly shorten the period of the initial reactions. The faster this heat is released, the higher the early strength, so that the 'critical strength' required to avoid frost damage is attained sooner. Thus, the heat released during the setting of the cement is of great practical value in cold weather concreting. The acceleration of strength and set time provided by accelerators is widely used by precast producers and floor finishing contractors to ensure that desired production and job schedules are not affected.

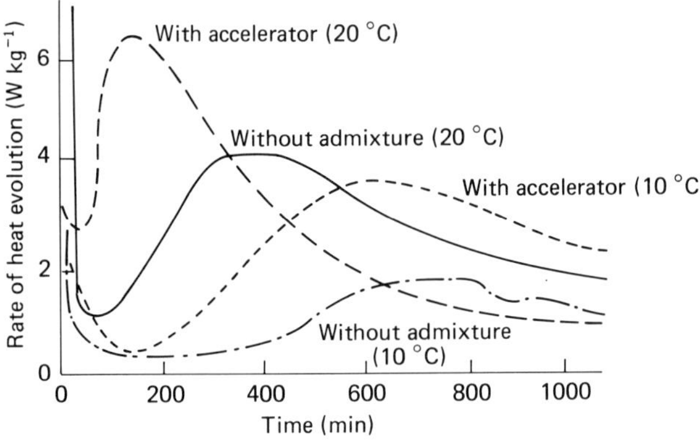

Fig. 5.35 Influence of low temperature and accelerating admixture on the heat of hydration curve.

5.8.2 Reduction of freezable water

The duration of protection required to protect concrete from frost damage will depend on a number of factors. One such factor is the degree of saturation of the freshly placed concrete, and the duration of the saturated state. Under normal conditions the degree of saturation is reduced by both the consumption of water in the hydration process, and the evaporation of the free water. Therefore, any factor which reduces the degree of saturation below the level which would cause damage by freezing will benefit cold weather concreting. Normal water-reducing and superplasticizing admixtures offer such benefit and this aspect is discussed below:

Approximately 65–70% of the water present in a concrete mix provides the workability required for placing operations. Such water is usually present as physically bound and free water filling the pores. If the required plasticity can be achieved with a much reduced water content, then the degree of saturation and vulnerability to frost attack is lowered. The 10–20% water reduction afforded by conventional water reducers and superplasticizers, therefore, offers a significant advantage in winter concreting.

Normal water reducers, due to the wetting action, increase the rate at which water combines with the cement particles. Superplasticizers are particularly noted for the high deflocculation they produce. Consequently, a good amount of the water used in the mix is quickly bound in the physical form of films round cement particles, reducing the free water.

5.9 Economic aspects

In estimating the economy that ensues by the use of an admixture, both changes in mix composition of a unit volume of concrete and cost of handling, transporting, placing and finishing the concrete should be taken into account.

Economies in mix design can be obtained through the use of both conventional and superplasticizing admixtures. The areas of high strength concrete and concreting in hot, arid climates offer the widest scope in this respect, since high cement contents and low water–cement ratios are specified in both situations.

Conventional lignosulphonate, carbohydrate and hydroxycarboxylic acid based water reducers permit 7–10% cement reductions in mixes with cement contents greater than 300 kg m^{-3} or in large scale construction with no deleterious effect on the finished product [68]. Superplasticizers produce dramatic water reduction (> 20%) at higher admixture dosages and will, therefore, afford greater cement reduction, or a switch from a more finely ground cement (Type III) to a coarser particle cement (Type I). This is shown in Table 5.7 for mixes with cement contents exceeding 400 kg m^{-3} [69].

Frequently, an admixture will allow the use of less expensive construction methods, a reduction of the numbers in a placing crew, or a reduction in the time required to carry out an operation, e.g. the placement of large volumes of concrete over extended periods minimizes the need for forming, placing and joining of separate units. In this respect, the use of flowing concrete in slab construction permits cost savings from a reduction in the number of operators in the placing crew or, alternatively, a reduction in the operational time when using the same number in the crew [70].

The process of precast concrete manufacture provides conditions which are more amenable to monitoring the cost benefit afforded by the use of admixtures. The following benefits have been found:

(a) High early strengths at reduced cost are attained so that forms are stripped earlier and rapid delivery time is ensured.
(b) Special concreting techniques may be used which provide suitable consistencies to place thin, highly reinforced structural elements.
(c) Lower curing cost.

Early strength is very important in precasting operations because, to a large extent, it controls the production cycle and the ability to accept and produce new work. This is particularly true of prestressed concrete where high early strengths are required prior to detensioning operations. Customary methods of assuring adequate strengths (about 30 MPa) have consisted of using excessively high cement contents resulting in concretes having 28-day strengths far in excess of that required for structural or durability reasons. Consequently, water-reducing admixtures are widely used to obtain the requisite early strength, but at reduced cement contents. This results in more economical mixes. Figure 5.36 shows a typical precast concrete yard specializing in prestressed products where water-reducing admixtures are used for this purpose. Typical results that can be obtained by the use of hydroxycarboxylic acid based normal water-reducing admixtures are presented in Table 5.16.

Fig. 5.36 A typical precast yard where the benefits of admixtures are obtained.

Table 5.16 The use of a hydroxycarboxylic acid based water-reducing agent to produce a more economical high strength mix for pre-stressed, pre-cast units

	Control mix	Modified mix
Mix composition		
RH Portland cement (kg)	285	255
Coarse aggregate 20 mm (kg)	830	860
Sand zone 2 (kg)	450	460
Added water (litre)	80	73
Hydroxycarboxylic acid water-reducing agent (ml/50 kg cement)	0	120
Concrete properties		
Slump (mm)	50	55
VeBe (s)	6.5	9.0
Compressive strength at: 1 day (steamed)	36.0	34.0
28 day	63.0	75.5

More substantial increases in the rate of strength development are obtained by the use of superplasticizers. As mentioned previously, because of dramatic water reduction that can be achieved by the use of higher dosages of superplasticizers, significant cement reduction or a change from a high early strength (Type III) to a normal Portland cement (Type I) is possible (Tables 5.17 and 5.7). The use of precast bridge beams where a naphthalene formaldehyde sulphonate based superplasticizer was used is shown in Fig. 5.37.

Superplasticizers may be used to realize significant energy savings in

Table 5.17 Application of high range water-reducing admixtures to reduction of mix cement content (Weston Hestor [71])

Slump (mm)	Accelerated cure compressive strengths (MPa) at age 16 to 17 hours		28-day compressive strength (MPa)	
	Mix 1	Mix 2	Mix 1	Mix 2
216	23.8	33.2	34.5	47.2
	23.8	32.5	33.2	46.5
89	32.4	31.8	36.6	41.4
	31.8	32.2	35.6	41.9

Accelerated curing: 3 hour precure 4.5 °C rise and 58 °C for 16–17 hours, and subsequently moist cured.
Mix 1: Cement content 351.9 kg m^{-3} type III cement, conventional carbohydrate type admixture at recommended dosage.
Mix 2: Cement content 286.6 kg m^{-3} type III cement, recommended dosage of sulphonated naphthalene formaldehyde superplasticizer.

Note: Anomolous decrease in strength values for specimens from concrete mixes of lower slumps attributable to mix thixotropy. Identical minimal consolidation was used for both high and low slump mixes.

Fig. 5.37 Precast bridge beams for highway spur in Italy. (Photograph – Courtesy, MAC Teviso Italy)

accelerated curing or even eliminate the necessity for curing [71]. High early strengths can be attained at considerably lower temperatures and curing times than are currently used. Figure 5.38 illustrates the advantage by showing a comparison of two mixes, one containing a superplasticizer and the other a plain mix, both steam cured. Detensioning strengths for the superplasticized concrete can be achieved in six hours of curing as opposed to the 22 hours curing regime required for the plain concrete.

Superplasticizers can be used successfully to achieve savings in concrete consolidation in heavily reinforced precast sections, e.g. the high quality

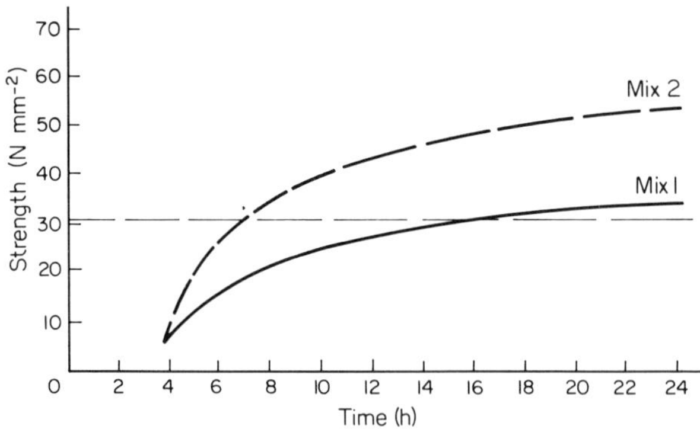

Fig. 5.38 Strength increases of cubes steamed under a normal heat profile showing that the destressing strength of about 30 N m^{-2} is obtained in 6–7 hours as opposed to 22 hours in the control mix.

Fig. 5.39 General view of the high density steel reinforcement used in the tube element of the Olympic Stadium in Montreal, Canada.
(Photograph - Courtesy, S.K.W. Trostberg, W. Germany)

finish required for the precast tube elements (carrying the mechanical and electrical services) used in construction of the Montreal Olympic Stadium was achieved through the use of a superplasticizer. Figure 5.39 shows the high density of the steel reinforcement use in these structures.

Concrete products like bricks and blocks which are mass produced can also benefit from the use of air-entraining and water-reducing agents particularly of the lignosulphonate type. These admixtures allow reductions in cement content of up to 10% without altering the mechanical properties of the product. Unlike normal concrete, the strength of concrete bricks and blocks is determined by the degree of interlocking the aggregates attain and is, therefore, more a function of compaction rather than cement content.

The cement paste is a discontinous phase that is present at points where the aggregates are in contact. In addition, on demoulding the blocks from the machine, a considerable degree of green strength is required which is also contributed by the cement. Reductions in cement content of up to 10% in the presence of a water-reducing admixture allow blocks to be produced and handled in the green state as well as having the required mechanical properties. Figure 5.40 illustrates the type of bonding in concrete blocks [72]. The ideal requirements for the paste component of the block are that it should be as fluid as possible to allow the maximum compaction and yet be stiff enough to maintain shape on demoulding and give maximum strength. It is found that this compromise can be obtained by incorporating water-reducing admixtures into the mix. Typical mix designs and resultant properties of the blocks produced on an automatic machine are given in Table 5.18.

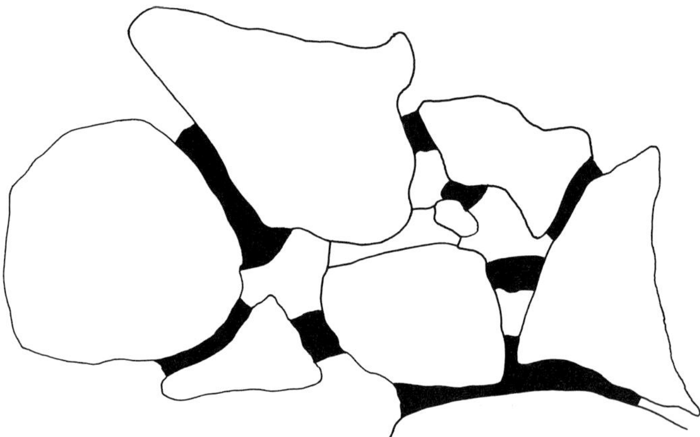

Fig. 5.40 The bonding of aggregates in blocks (McIntosh).

Table 5.18 A lignosulphonate based water-reducing agent allows lower cement contents to be used in concrete blocks whilst maintaining the compressive strength

Mix composition					
R H Portland cement (kg)	90	83		80	
Zone 2 sand (kg)	550	550		550	
Limestone aggregate (kg)	650	650		650	
A/c ratio	13.3 : 1	14.5 : 1		15.0 : 1	
Admixture (lignosulphonate at 140 ml/5 kg cement)	No	No	Yes	No	Yes
Compressive strength of blocks					
(N mm^{-2}) at: 33 h	2.5	2.4	2.7	2.3	2.5
28 day	6.0	5.4	6.5	4.8	6.5

Similar factors are applicable to extruded products where normal and retarding water-reducing admixtures enhance lubricity and cohesion of the mix allowing easy extrusion and yet maintaining the product profile on emergence from the dye.

The economics of block production, particularly the 'egg laying' process, is related to a large extent to the length of time before the blocks can be lifted and stacked. Ideally this should be carried out the day after production, but in winter conditions the strength attained is often inadequate to permit this. The use of accelerating admixtures to achieve the required strength is commonplace and, indeed, in certain cases accelerating water-reducing admixtures are used simultaneously to reduce the cement content.

Admixtures of various chemical classes provide a means to control and improve the quality of concrete, facilitate design and construction techniques not readily practicable with concrete not containing such materials. They should not be substitutes for poor concrete materials or poor concreting practices, but used as beneficial components of concrete which, when properly employed, broaden the applications of concrete.

5.10 General guidelines for the use of admixtures in concrete

The information presented in this section covers the operational aspects pertaining to the use of admixtures in the field. It also offers guidance in the selection, use and control of uniformity of the product. Most of the items discussed here apply equally to the three major fields of concrete construction, namely, site batched and placed concrete, ready-mixed concrete, and precast concrete.

5.10.1 Preparation

Since the quantity of an admixture used is small relative to other contituents of the mix and its effect is proportionate to the amount used, it is apparent that the method used to prepare and measure the admixture is important.

Most of the proprietary admixtures available in the market-place are ready to use liquids which require no further preparation at the location of use. Indeed, site preparation with these products is strongly discouraged by the manufacturer. There are, however, exceptions to this rule, as given below:

(a) Admixtures are sometimes supplied in concentrated form to minimize high freight cost that would otherwise be incurred in the transport of significant amounts of water.
(b) Calcium chloride flake is dissolved in water at the plant by some concrete producers.

(c) Chemical admixtures are sometimes supplied as water soluble solids requiring mixing at the job site or point of use.

Preparation may involve making standard solutions (as in the case of calcium chloride and powder admixtures) or dilution to provide the manufacturer's recommended dosage or facilitate accurate dispersion [67]. A number of problems may arise in the preparation of admixtures such as difficulty in mixing and sedimentation of insoluble materials or active ingredients. It is, therefore, good practice to seek the manufacturer's advice if difficulties arise.

All mixing of powdered admixture must be done in a mixing tank which is separate from the admixture storage tank from which the admixture is fed into the dispenser. The density of the admixture mixed in this manner should be checked daily or when new material is mixed, using hydrometers. The storage tanks should be equipped with agitators to keep the admixture in suspension.

5.10.2 Containers and delivery

Various forms of packaging for admixtures are available (Fig. 5.41).

Liquids

The majority of proprietary admixture compositions are neutral or slightly alkaline solutions that are supplied in:

(a) Five or twenty-five litre plastic containers made from low or high density polyethylene or polypropylene.
(b) 210 litre lacquer-lined mild steel drums: second-hand relined drums are quite satisfactory and are, therefore, commonly used. It is useful to have such drums in a suitable cradle and a small opening should be available for fitting a tap in addition to the normal filling bung.
(c) 1000 litre portable tanks: constructed from mild steel or heavy duty polypropylene with a steel frame have been found suitable. The tanks are useful for medium sized precast and site situations and can be delivered filled on a flat truck. Handling can be done be a crane or fork lift truck.
(d) Bulk storage tanks: cylindrical or rectangular shaped tanks ranging in capacity from 2500–15000 litres are available. Although plastic tanks are sometimes used, those constructed of mild steel are more common. The required fittings include delivery and ventilation pipes and an outlet pipe complete with a tap a few inches from the bottom. The tank should be clearly marked with a label that cannot be altered and situated in a conspicuous area such as near the access opening. Safety and handling

Fig. 5.41 The various forms of packages commonly used for the delivery of admixtures: (a) 5 and 25 litre containers; (b) 210 litre lined drums; (c) 1000 litre portable tanks; (d) 25 kg multi-ply sacks.

aspects in a printed form should always be available. Delivery to fixed tanks is made from road tankers or occasionally 20-tonne rail tankers. Specialized vehicles containing several compartments with individually calibrated meters with print-out facilities are finding more widespread use (Fig. 5.42). This latter method of delivery is particularly useful since it removes the necessity for inaccurate tank dipping or vehicle weighing before and after deliveries. It is required by law that such tankers should carry advice on hazards presented by accidental spillage of materials being carried.

Solids

Admixtures sold in powder form are usually packed in either multi-ply paper sacks or in small fibrite kegs. Typical powdered materials are waterproofing

Fig. 5.42 Speciality admixture delivery vehicle.

admixtures based on stearates or stearic acid, accelerators like calcium formate, some air-entraining agents and the spray dried solid forms of superplasticizer.

Labels

All receptacles in which admixtures are delivered or stored should be clearly labelled. The standard information which must be featured is as follows:

(a) Name, address and telephone number of producer and distributor.
(b) The trade designation of the product, i.e. brand name, reference number and/or letter.
(c) Designated type as given by the principal and secondary functions, e.g. accelerating, water-reducing or air-entraining water-reducing admixture.
(d) Basic instructions for use and precautions where necessary regarding handling of the admixture, e.g. safety aspects such as caustic, toxic or corrosive characteristics.
(e) Information regarding the presence of chlorides. The chloride content should be stated and also expressed as a percentage (of anhydrous chloride ion) of cement at the recommended dosage [67].
(f) Any special storage or shelf-life data.
(g) The relevant standards (ASTM, BS, CSA) number.

(h) The recommended dosage or dosages and the maximum amount not to be exceeded. The maximum amount should not be exceeded or secondary effects (such as excess air entraining and set retardation) can occur.

The above information should be placed within the frame of the label and marked in such a way that the information connot be altered. An example of a typical label is shown in Fig. 5.43.

Fig. 5.43 A typical label for admixtures.

5.10.3 Safety aspects

The supply of information necessary to ensure the health and safety of personnel handling admixtures is mandatory. Such information is usually contained in the hazard data sheet required by the national organization for health and safety.

The information required includes the following:

(a) Nature of the hazard, e.g. caustic, inflammable, etc.
(b) Advice on protective clothing, e.g. gloves, goggles, boots etc.
(c) Action to be taken in the event of skin or eye contact.
(d) Required action in the event of spillage, e.g. type of extinguisher if inflammable, washing of slippery floors, etc.
(e) Toxicity and action in the event of ingestion or prolonged exposure.

5.10.4 Storage

Most admixtures, with the exception of some waterproofers, non-chloride accelerators and air-entraining agents, are aqueous solutions which freeze at about −3 °C. Typical data for minimum storage temperatures for the various types of admixture currently marketed are presented in Table 5.19.

Table 5.19 Approximate minimum storage temperatures for various admixtures before separation or solidification occurs

Category	Type	Approximate minimum storage temperature (°C)
Water-reducing	Lignosulphonates	−3 to 0
	Hydroxycarboxylic acid	−5 to 0
	Hydroxylated polymer	−5 to 0
	Superplasticizers	−3 to 0
Air-entraining	Neutralized wood resins	−3 to 0
	Fatty acid soaps	
	Alkyl-aryl sulphonates	
Waterproofers	Fatty acid emulsions	1
	Wax emulsions	
Accelerators	Calcium chloride solution	−10 to −5

For the majority of materials in UK winter conditions, storage in drums in unheated enclosed buildings is sufficient to prevent freezing. In North America, however, where winters are much more severe, adequate provision should be made to prevent freezing of liquid admixtures. Protection must refer to the whole system, which include storage and mixing areas, lines to the dispenser and lines to the mixer. Bulk storage in steel or plastic tanks should be inside a heated and enclosed building, although outdoor lagged tanks have been found satisfactory in the milder climate of the UK and some coastal and southern parts of North America.

Most of the frozen admixtures mentioned in Table 5.19 may be used in concrete after thawing and thorough remixing. However, some of the emulsions will not reconstitute on thawing and will, therefore, require greater care in storage to prevent freezing. The thawing of frozen admixtures can be done by bringing the materials into conditions at least 10° C above their freezing temperature and maintaining this temperature for a period of time. The content of the drum or tank should then be agitated, either manually or using a low pressure air sparge to produce a homogeneous solution or emulsion. It should be emphasized that frozen admixtures require both thawing and remixing prior to use. The use of admixtures which have only been thawed with no remixing may result in severe retardation of concrete. Clogging of lines and dispenser may also occur because of separation of the admixture into a supernatant water layer and a thick lower liquid that occurs on thawing.

Solid admixtures are not generally affected by normal winter conditions, but should be stored in areas free from dampness, condensation or rain, preferably in an elevated location. The materials in ripped bags should be used as soon as possible after the bags are damaged or should be discarded.

Elevated temperatures, even those prevailing in tropical climates, will not significantly alter the effectiveness or shelf life of the majority of admixtures with the following reservations: (a) Impure lignosulphonates, hydroxycarboxylic acids, hydroxylated polymers and fatty acid based materials can support fungal and bacterial growth. Deterioration of the materials due to bacterial and fungal activity is usually accelerated at elevated temperatures, and suitable materials (e.g. formaldehyde or sodium-*o*-phenyl phenol tetrahydrate) are normally added to prevent this from happening. Potential problem regions are hot weather countries and parts of North America which get hot during summer months. Admixture suppliers in these locations should, therefore, always take the required precaution to discourage such microbial and fungal activity. (b) Emulsified products, such as waxes or fatty acid waterproofers, become unstable at elevated temperatures and may be difficult to reconstitute on lowering the temperature. Advice should be sought from individual manufacturers, but the use of such materials is not advised in hot climates.

5.10.5 Batching

(a) Point of addition

The rate of discharge, timing in the batching sequence and the amount of material used are primary criteria to be considered in the incorporation of admixtures into a concrete batch. Thus, a procedure for controlling the time and rate of addition to the concrete batch should be established and closely followed. Additionally, the compatibility of the admixtures intended to be used in the same concrete batch when discharged in the same water phase should be determined. In general, most manufacturers discourage the intermixing of admixtures prior to introduction into the mix. When two or more admixtures are to be used in the mix, they should be incorporated at different times during the mixing. If separate addition is impossible, then the manufacturer's advice should be sought on the matter. Table 5.20 summarizes the preferred points of admixture addition in the mixing cycle to obtain best results.

In situations where rapid repetitive mixing occurs, it is often not possible to adhere closely to these recommendations; it is particularly important in

Table 5.20 Preferred point of addition in the mixing cycle for different admixture types

Admixture type	Point of addition	Notes
(a) All water-reducing admixtures (except superplasticizers used for flowing concrete production) particularly when a retarding effect is required	After initial mixing period of up to 30 s of aggregates, cement and part of the gauging water. Should be dissolved in a proportion of the remainder of the gauging water	Pre-mixing with moist aggregate is often sufficient to partially hydrate the cement
(b) Air-entraining agents (c) Accelerators (except powders) (d) Emulsified water-proofers	Dissolved in gauging water which can be added directly to the pre-mixed cement and aggregates	These materials are not particularly sensitive to the point of addition and the main ambition is to obtain the maximum dispersion through the mix
(e) Powdered waterproofers (f) Powdered accelerators	Pre-mixed with dry mix ingredients before addition of the gauging water	In order to aid the dispersion of this type of mix it is advisable to sprinkle the admixture over the mix during the dry cycle
(g) Superplasticizer for flowing concrete	After the mixing cycle just prior to placing	

the case of water-reducing retarding admixtures where the difference between adding the material in the initial gauging water and adding it after a premixing period can be two to three times the normal retardation time.

(b) Batching equipment

For the proper use of chemical admixtures in concrete mixes, a means must be provided for accurate and reliable batching of the admixtures into the concrete batch. Admixture batching systems are two types:

(i) Liquid batching systems for materials introduced in the liquid form.
(ii) Dry batching systems for materials which are powders.

(i) *Liquid batching systems.* Liquid admixtures are usually batched by dispensers. A great variety of dispensing systems are used including simple pumps, timers, visual volumetric containers, positive volumetric displacement systems and weight batching. The complexity and expense of the system used is largely dictated by the size or value of the particular application. However, the basic requirements of all methods are an accuracy of ± 3% [73–77] and an insensitivity to changes of viscosity or temperature. The device should be constructed with materials which will not be corroded by the admixture being used. All components should be of proven low failure rate to ensure reliability in the long term. Additionally, Concrete Plant Manufacturer's Bureau (CPMB) standards require that provision be made available on each dispensing device for routine diversion of a measured dosage into a small container for verification of the batch quantity. Specific requirements pertaining to dispensing systems are

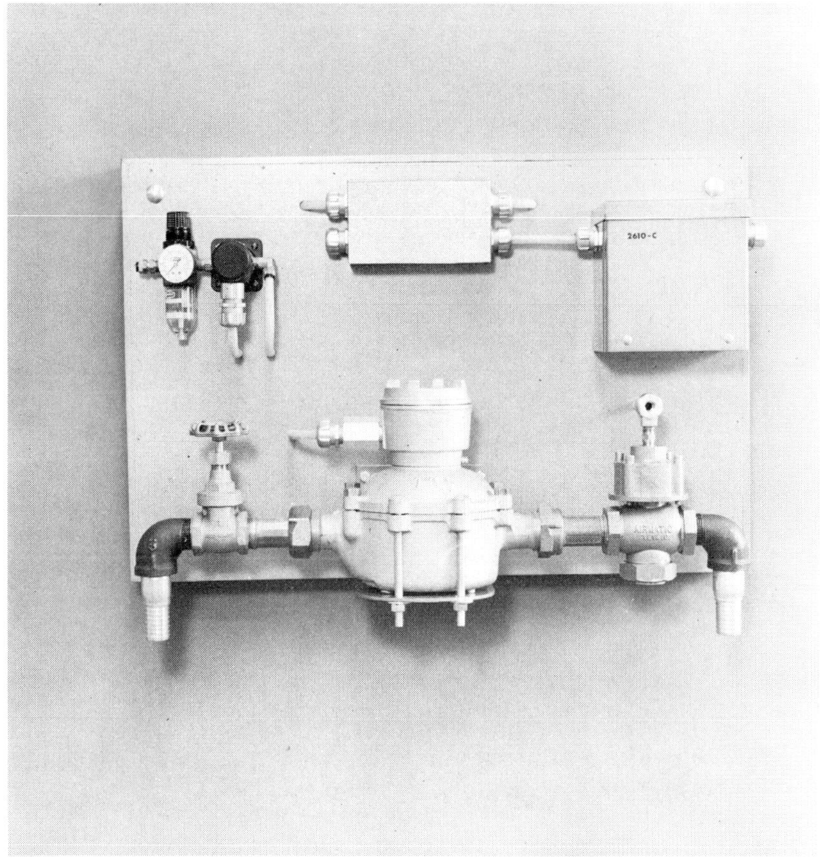

Fig. 5.44 Typical positive displacement flowmeter used for variable admixture dosage. (Photograph – Courtesy, Protex Industries Inc., Denver, Colorado, USA)

detailed in BS 5328, ASTM Publication C-94, NRMCA Check List, CPMB Standards and the NBS Handbook.

Essentially a liquid batching system will include a method of moving the liquid from the feed tank to volumetric containers (usually by pneumatic or electric pumps) a batching or dispensing device, controls and appropriate interlocking devices. Visual volumetric containers are required for all liquid systems.

Measuring devices are either positive volumetric displacement devices or timer controlled systems. The former include flowmeters and measuring containers equipped with floats or probes. Flowmeters may be equipped with pulse emitting equipment which operates a preset electrical counter on the control unit located near the batching console [67]. Such meters are usually geared in a manner which facilitates variable admixture dosage. Figure 5.44 shows a typical displacement flowmeter operated by pneumatic power, while Fig. 5.45 shows a control unit on which the required dosage of admixture can be preset.

A measuring container operates by the principle whereby the linear movement of a float in a container of given cross section meters out a volume

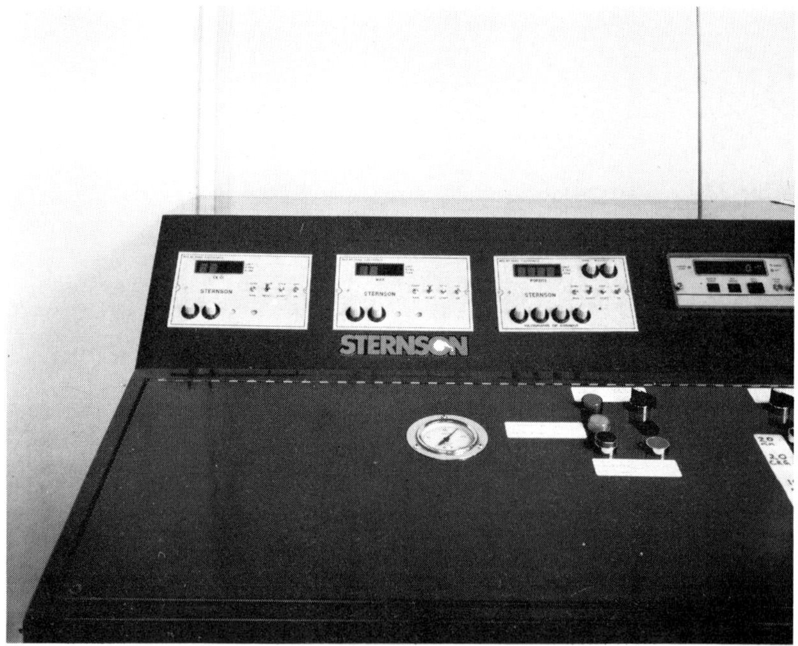

Fig. 5.45 Control unit used for the control of liquid admixture dosage in dispensing system. (Photograph – Courtesy, Sternson Ltd., Australia)

of solution required for the batch. Usually the floats are connected to a potentiometer or pulsating switches which operate electrical preset counters [67]. A typical unit is shown in Fig. 5.46 where it can be seen that the measuring device consists of a sight glass containing a moveable electrode, which in this case is connected to a dial calibrated in cement content. The admixture is pumped into the sight glass until it touches the upper electrode when the pump is automatically switched off and the material is allowed to discharge into the water pipe.

Fig. 5.46 Electrically operated dispensing equipment.

Timer controlled systems involve the timing of flow through an orifice and have been successfully used with high volume dilution admixtures [67]. Most systems include a sight glass or other means of checking the amount batched. However, due to the number of variables that can introduce error into the measurement, such as changes in viscosity of the admixture or in the power supply, their use is generally discouraged [78]. Timer systems are permitted only for calcium chloride in the NRMCA Check List. The advantages of the system is that it is simple and basically of low cost. Figures 5.47 and 5.48

Fig. 5.47 A timer control unit. (Photograph – Courtesy, D. Lamb, Master Builders Co., Toronto, Canada)

show a timer control unit and the unit installed with visual volumetric containers respectively.

Weight batching of admixtures using beam and dial scales with an indicator in the batching system to register discharge of the material is occasionally used to measure the addition of admixtures. The disadvantages of this system include the necessity for conversion of the admixture dosage from volume to weight, and the necessity to dilute admixture solutions to obtain a sufficient quantity for accurate batching. The method has been found suitable for air-entraining admixtures.

In general, most dispensing systems are custom-made to meet the method of control or degree of automation required for the particular operation. Batching systems and controls can therefore be classified as fully automated, semi-automated or manual systems.

(1) *Fully automated dispenser.* This is the preferred method of adding liquid admixtures to a concrete and is based on either electric or pneumatic systems. Such a system consists of the required combination of individual volumetric devices, all of which are automated. The system should also have the following features [79]:

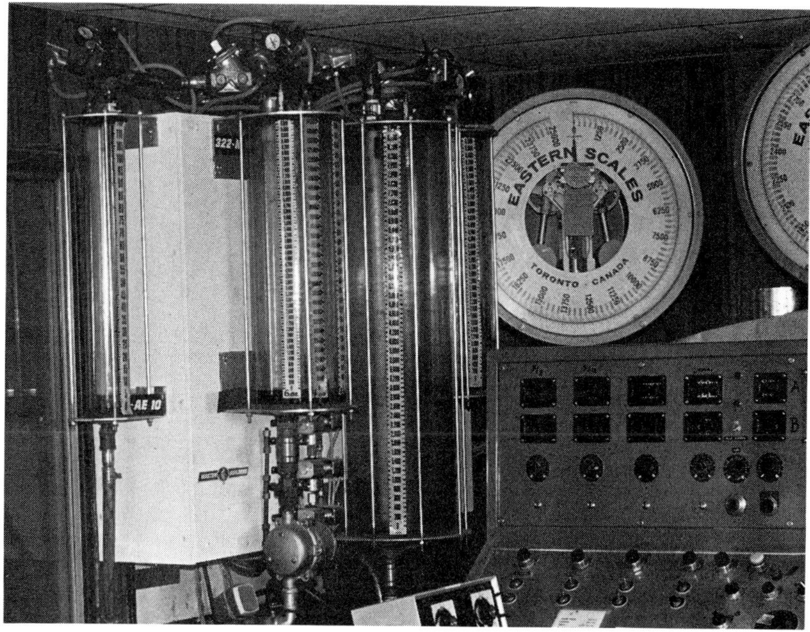

Fig. 5.48 Timer controlled unit installed in a ready-mix concrete plant. (Photograph – Courtesy, D. Lamb, Master Builders Co., Toronto, Canada)

(i) A single starting mechanism or separate starting mechanisms for admixtures not batched at the same time as other ingredients.
(ii) Signalling of empty and resetting to start by volumetric devices.
(iii) Provision of a method which prevents the discharge of any ingredient until all individual batches have been cleared of the previous batch and returned to zero within tolerance.

This type of dispenser is shown installed in a ready-mix concrete plant in Fig. 5.49. It is possible to operate the control release automatically so that when the batching console is used in the normal way a signal is picked up by the dispenser relay and the material is automatically fed into the mixer.

(2) *Semi-automated dispenser.* This type of dispenser is widely used and can operate either electrically or pneumatically. The system contains the necessary individual volumetric devices, the controls of which are either semi-automatic, interlocked, or automatic or a combination of these. With this type of equipment the level of admixture is allowed to rise in the sight glass by the operation of a switch or valves until a correct level is reached. The valve is then opened to allow dispensing of the admixture into the water

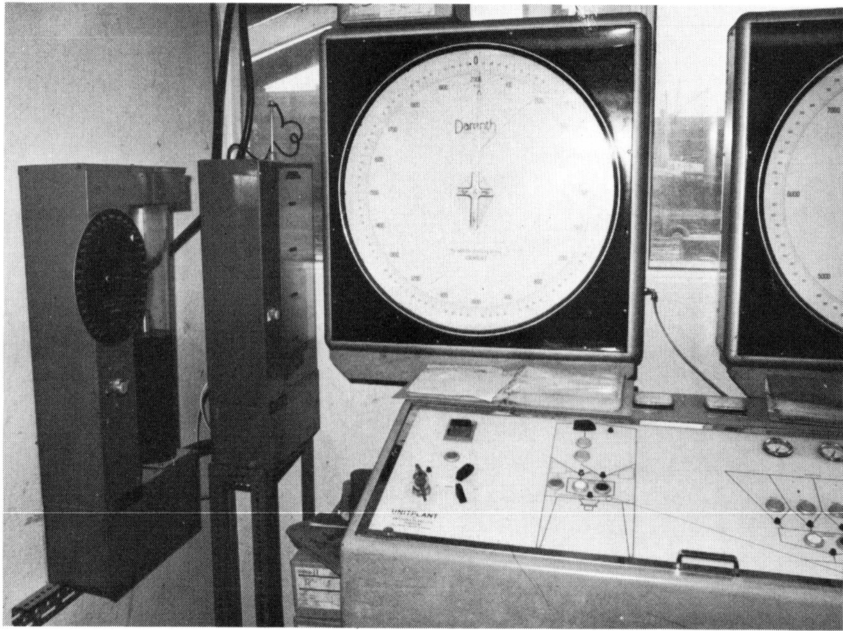

Fig. 5.49 Electrically operated fully automatic dispenser.

pipe or mixer. An illustration of this type of pneumatically operated dispenser is shown in Fig. 5.50.

The arrangement of the piping system in dispensing equipment requires careful attention to ensure that variable amounts of material are not trapped in the lines, that valves will not leak and the amount of material measured is actually discharged into each batch. [79].

(3) *Hand pump.* This type of transfer measuring device is suitable for small operations where admixtures are only occasionally used. It is normally constructed to fit into a 210 litre drum bung hole. Two types are available: one is a simple pump which will require hand dispensers of the types described below to measure the actual quantity; the other is a type in which one stroke of the pump delivers a set volume. It can be used directly for adding the admixture to the concrete mix.

(4) *Hand dispensers.* Containers of standard and correct volume may be used in small applications or for special mixes. It is important that a dispenser of the correct size for one mix should be used rather than a measuring cylinder or other partly filled vessel.

(ii) *Dry batching systems.* The metering of powdered admixtures such as

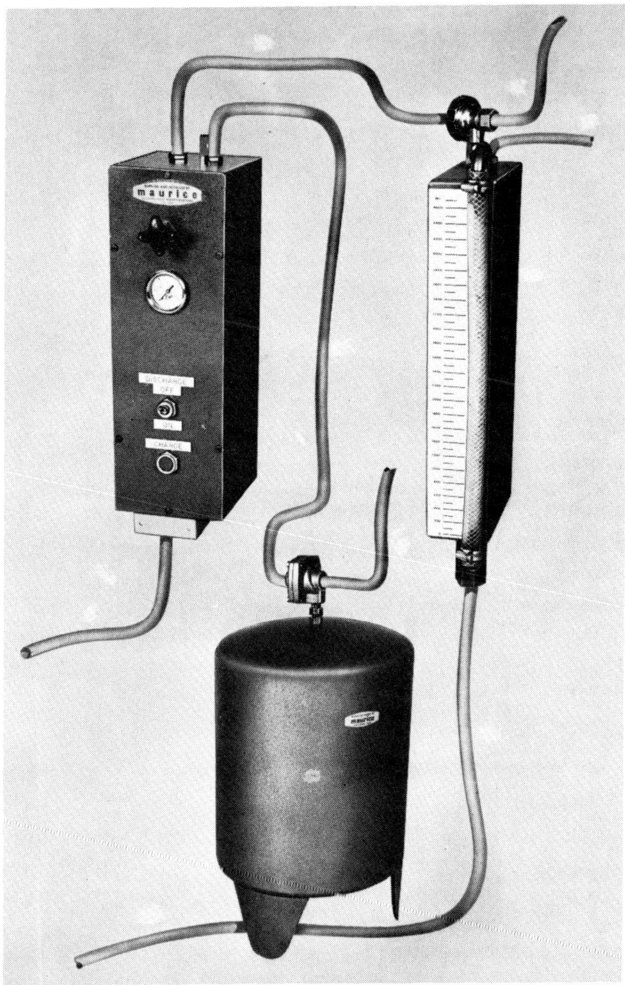

Fig. 5.50 A pneumatically operated semi-automated dispenser.

calcium formate or some waterproofers can be readily done by weight. When relatively small quantities of such admixtures are used it is wise to blend them with finely divided inert or pozzolanic material. This not only facilitates more accurate weighing but also prevents flotation of the powder on the mix water thus ensuring better dispersion in the mix. In the use of powdered admixtures care must be taken to prevent the adherence of the powder to the sides of the mixer or the formation of clumps. These problems can be averted if the powder admixture is not dispensed with the mixing water. A suitable point of addition in the mixing sequence is after the fine aggregate has been added.

A suitable measuring device for powders is the automatic pigment scale. Most of the units commercially available consist of a part that operates volumetrically (e.g. screw, shaking gutter, etc.) and a suitable metering scale. Space is essential in positioning the metering unit and powder admixture containers as closely as possible to the mixer or raw material supply lines. Care must be taken to ensure that the powder that enters the concrete mixer is free flowing [80]. These automatic scales are quite expensive.

(c) Evaluation and selection

Admixtures should be purchased under specifications that stipulate the desired properties, exclude adverse effects and permit evaluation of the uniformity of the admixture from batch to batch. The requirements for admixtures should be based on performance tests as stipulated by the recognized relevant standard specifications. However, specifications deal primarily with the influence of admixtures on standard properties of fresh and hardened concrete. The potential user may be interested in other properties of the concrete, particularly enhanced workability, placing and finishing qualities, and early strength development. These secondary functions may, in some instances, be of greater importance when the selection of admixture is being considered and also in the determination of the intended dosage.

The tests detailed in national standard specifications are, therefore, essentially qualifying tests which afford a screening procedure in the selection process. Suitable admixtures should then be tested with job site concrete materials under actual operating conditions, so that a proper measurement of the desired engineering properties is obtained. This will permit adjustment of the manufacturer's recommended dosage that may be necessary to enhance a particular secondary effect. Finally the cost of the concrete containing the admixture should be established.

(d) Admixture uniformity

Any substantial use of admixtures in continuing production of concrete should be accompanied by routine sampling and testing of the admixture to determine within-lot uniformity of the product and also to provide data showing that any lot is the same as that previously supplied. For the most part, this will be based on physical and chemical index tests, e.g. solids content, specific gravity of solutions and infra-red spectroscopy of active constituents. Such tests, although not definitive of the quality of the product, are usually determined mainly to assure uniformity of the product being supplied.

(e) Precautions

The manufacturer's instructions should be carefully followed when using admixtures. The specific effects produced in concrete by admixtures are dependent on a number of factors such as cement composition, aggregate characteristics, the presence of other admixtures and ambient conditions. The full potential of admixtures in both economy and desired engineering objectives can often be realized only by changes in the proportioning of the concrete mix otherwise employed. Therefore, it is important that laboratory and field trials be carried out to select the appropriate admixture type and dosage to be used. The effects obtained with the admixture using job materials under job site conditions should be determined. This is particularly important when special types of cements are specified, when more than one admixture is to be used, or when mixing and placing is carried out at ambient temperature conditions well above or below generally recommended concreting temperatures [67]. Care should be exercised to ensure that the admixture is properly and uniformly dispensed.

References

1 Gordon, W.A. (1966). *Freezing and Thawing of Concrete, Mechanisms and Control.* ACI Monograph No. 3.
2 *Design and Control of Concrete Mixtures* (1978). Engineering Bulletin of the Portland Cement Association, Skokie, Illinois, USA.
3 Madderom, W.F. (1980). *Concrete International* February, 110–114.
4 Hyland, E.J. (1982). Private Communication.
5 Taylor, T.G. (1948). *ACI Journal,* Proceedings **45,** 469–87.
6 Kobayashi, M. (1967). *Proceedings of the International Symposium on Admixtures for Mortar and Concrete,* Brussels, 87.
7 Malhotra, V.M. (1981). *Report MRP/MSL 81–53 (J),* Draft CANMET, Ottawa, Canada.
8 Blanks, R. F. and Gordon, W. A. (1949). *ACI Journal,* Proceedings **45,** 489–97.
9 Malhotra, V.M. and Malanka, D. (1977). CANMET Report, Department of Energy, Mines and Resources, Ottawa, Canada, 65–77.
10 Mielenz, R.C. and Sprouse, J.H. (1979). *Superplasticizer in Concrete.* ACI Publication, SP-62, 171– 92.
11 Mather, B. (1978). *Superplasticizer in Concrete.* ACI Publication, SP-62, 158–66.
12 Hester, T.W. (1979). *Superplasticizers in ready-mixed concrete. (A practical treatment for everyday operations).* Publication No. 158, NRMCA, Silver Springs, Maryland, USA.
13 Curtis, R.J. (1975). *Contract Journal,* **10,** 24–5.
14 Newman, K. (1975). *Proceedings of the Conference on Ready-Mixed Concrete.* Dundee, Scotland.

15 Howard, E.L., Griffiths. K.K. and Moulton. W.E. (1959). *Journal of Materials,* **14,** 220–230.
16 Freedman, S. (1970). *Modern Concrete,* November, 170–76.
17 Aitcin, C.P. (March 1980). *Concrete Construction,* **6,** 220–30.
18 Anon (1962). Army Engineering Waterways Experiment Station, Mississippi. AD-756299.
19 Fitzgibbon, M.E. (1975). *Proceedings of the CS Symposium 'Tomorrows Concrete'.* Doncaster, UK.
20 Browne, R.D. (1973). *Proceedings of the Conference on Large Pours for RC Structures.* University of Birmingham, 44–8.
21 Forbrich, L.R. (1940). *ACI Journal* Proceedings, **37,** 161–83.
22 Morgan, H.D. (1958). International Commission on Large Dams, *Question No. 23,* **No. R-12,** 10–24.
23 Shutz, R.J. (1959). *ACI Journal,* **11,** 769–81.
24 Mather, B. (1982). *Concrete International,* **4,** 112–18.
25 Gerwick, J.R. C.B. (1980). *Research requirements for concrete in marine environments.* ACI SP-65, Symposium on Performance of Concrete in Marine Environment, St. Andrews by the Sea, Canada, 577–88.
26 Mailvaganam, N.P., Bhagrath, R.S. and Shaw, K.L. (1983). *Effects of chloride and non-chloride admixtures on superplasticized silica fume concrete.* Annual Meeting of the Transportation Research Board, Washington, D.C.
27 Fisher, C.H. (1972). *ACI Journal,* **69,** 556–61.
28 O'Brien. J. (1973). Cement and Concrete Association of Australia, Technical Report No. 33.
29 Valore, JR. R.C. (1978). *Significance of Tests and Properties of Concrete and Concrete Making Materials.* **STP 169 B,** ASTM, 860–72.
30 Anon (1968). *Concrete Construction,* December, 860–72.
31 Gerwick, Jr. B.C., Holland, T.C. and Komendant, G.J. (1981). *Tremie Concrete for Bridge Piers and Other Massive Underwater Placements.* Report No. FHWA/RD–81/153, Federal Highway Adminstration, US Department of Transportation, Washington, D.C.
32 Tynes. W. (1967). *Evaluation of admixtures for use in concrete to be placed under water.* Technical Report C-67-3, US Army Engineer Waterways Experiment Station, Vicksburg, Mississippi, USA.
33 William, Jr. W. (1959). *ACI Journal* **9,** 362–66.
34 Anon (1968). *Concrete Construction,* July, 815–23.
35 *Tilt-Up Construction.* (1980). ACI Compilation No. 4, April, 51–5.
36 Valore, R.C., Kudrenski, W. and Gray, D.E. (1978). *First International Symposium on Superplasticizers in Concrete.* Ottawa, Canada. ACI Publication SP-62, 559–608.
37 Bonzel, J. and Siebel, E. (1975). *Fluid concrete and its possibilities of application.* Translation from *Beton* **24,** 20–4.
38 Freese, D. (1978). *First International Symposium on Superplasticizers in Concrete.* Ottawa, Canada, 133–60.
39 Diamond, C.P. and Bloomer, S. (1977). *Concrete,* December, 217–22.
40 Bartos, P. (1978). *Concrete,* October, 241–4.
41 Seabrook, P.T. and Malhotra, V.M. (1979). *Superplasticizers in Concrete.* ACI Publication SP-62, 263–92.
42 Lapinas, R. (1982). *Flowmix Canada.* Canada Cement Lafarge. Private Communication.
43 Previte, R.W. (1977). *ACI Journal,* Proceedings **124,** 361–6.
44 Ravina, D. (1975). *ACI Journal,* Proceedings **62,** 291–95.

45 Rixom, M.R. and Wadicor, J. (1981). *Developments in the Use of Superplasticizers* ACI Publication SP-68, 359–80.
46 Collepardi, M. (1976). *Cement and Concrete Research,* **6,** 401–7.
47 *Superplasticizing admixtures in concrete.* (1976). Joint Working Party Report, 45–030, Cement and Concrete Association and Cement Admixtures Association.
48 Bonzel, J. (1974). *Directives for the production and manufacturer of flowing concrete.* Translation from *Beton Herstellung Verwendung* **24,** 4.9.S. 342–4.
49 *Trial Mixes Using Mighty 150.* (1975). WRA Report No. 2 produced for Conforce Products Ltd, by R.M. Hardy Associates, Edmonton, Alberta.
50 Hester, W.T. (1978). *First International Symposium on Superplasticizers in Concrete.* Ottawa, Canada. ACI Publication SP-62, 533–58.
51 Khalil, S.M. and Ward, M.A. (1979). *Fourth International Symposium on Concrete Technology.* Monterrey, Mexico, 35.
52 *High Range Water-Reducer Committee Recommended Practice.* (1980). Draft of the Prestressed Concrete Institute.
53 Mailvaganam, N.P. (1978). ACI Publication SP-62, 389–404.
54 *Testing of Melment L-10 for Compatibility with Porzite Admixtures.* (1976). Laboratory Report, No. I/11/76, Sternson Ltd, Brantford, Canada.
55 Tynes. W.O. (1977). *Investigation of proprietary admixtures.* Technical Report C-77-1, US Army Waterways Experimental Station, Vicksburg, Mississippi.
56 Popovics, S. (1968). *Concrete* **4,** 272–6.
57 ACI Committee 201. (1977). *Guide to durable concrete.* (ACI 207 2R77), ACI, Detroit.
58 Rosenburg, A.M., Gaidis, J.M., Kossivas, T.G. and Previte, R.W. (1977). *A corrosion inhibitor formulated with calcium nitrite for use in reinforced concrete.* STP-629, ASTM, 228–40.
59 Daugherty, E.K. and Kowalewsky, Jr. M. (1967). *Journal of Materials,* **11,** 161–6.
60 Keeley, C. and Holdsworth, R. (1978). *Middle East Construction,* October, 104–7.
61 McCarthy, M. and Hodgkinson, L. (1978). *Admixtures for concrete in hot climates.* Technical Publication, Cormix Division, Joseph Crossfield & Sons Ltd.
62 Gayner, J.F.C. (1979). *Symposium on Concreting in Hot Weather Conditions.* Cement and Concrete Association.
63 McCarthy, M. (1979). *Symposium on Concreting in Hot Weather Conditions.* Cement and Concrete Association.
64 McCurrich, L.H. and Lammiman, S.A. (1979). *Symposium on Concreting in Hot Weather Conditions.* Cement and Concrete Association.
65 Perenchio, W., Whiting, D.A. and Kantro, D.L. (1979). *Superplasticizers in Concrete,* ACI Publication SP-62, 137–55.
66 Haddad, J.G. (1975). *Conditions for the Production, Placing and Curing of Winter Concrete.* ACI Chapter Meeting, Ottawa, Canada.
67 Admixtures for concrete. (1981). ACI Report 212-IR-81. *Concrete International,* 24–52.
68 Mielenz, R.C. (1960). *Symposium on the Effects of Water-Reducing and Set-Retarding Admixtures on the Properties of Concrete.* ASTM SP-266, 161–82.
69 La Fraugh, W.R. (1978). *Proceedings of the First International Symposium on Superplasticizers in Concrete.* Ottawa, Canada, 161–82.
70 Zummo, M.and Henry, L.R. (1981). *The use of high range water-reducing admixtures in residential concrete.* Melment Symposium, Munich.

71 Hester, W.T. (1978). *PCI Journal,* **23,** 68–85.
72 McIntosh, J.D. (1965). *Concrete Building and Building Products,* Vol. 2. 83–85, 89–97.
73 British Standard 5328 (1976).
74 *Concrete Plant Standards of the Concrete Plant and Manufacturers' Bureau* (1977). Sixth Revision, Silver Springs, Maryland, USA.
75 ASTM Standard Specification C-94 (1980).
76 NBS Handbook 44, National Bureau of Standards, USA.
77 *Certification of Ready-Mixed Concrete Production Facilities.* (1976). National Ready-Mixed Concrete Association, Silver Springs, Maryland, USA.
78 Parsons, J.S. (1973). *Report on Ready-Mixed Equipment.* Ministry of Transport and Communications, Ontario, Canada.
79 Gaynor, R.D. (1978). *Ready-Mixed Concrete,* Chapter 29, ASTM STP 169 B, 471–502.
80 Anon. (1983). *Metering of inorganic pigments in the manufacture of coloured concrete products.* Technical Information Leaflet, Mobay Chemical Corporation, Pittsburg, PA, USA.

6
Specifications and standards for chemical admixtures

6.1 Introduction

Many countries have introduced national standards or specifications for concrete admixtures or have alternatively set up some sort of control other than a national standard. The need for such standards can be summarized as follows:

(a) In the earlier years of the admixture industry and to some extent even today, there exists the problem of communicating across a chemist/engineer interface. This led to pressure for the establishment of some performance-related specification method, the development of which was the result of joint contributions from both sides of the industry.

(b) All other concreting ingredients (cement, aggregates, water, fly ash) were in most countries subject to some sort of standard specifications. As the use of chemical admixtures increased, so the need for some sort of conformity to specifications became more obvious.

(c) More recently, as the construction industry has increasingly turned to contracts outside its own particular national boundaries, the need to relate all materials to some known standard has been felt. Whilst the various concreting (and other) materials are brought in from all over the world, it is natural that the particular construction company would wish to assess the quality of these materials against its own national standard. It is partly due to this reasoning that there is currently a move towards the establishment of a DIN standard for admixtures in Germany.

6.2 The role of specifications and standards for chemical admixtures

The availability of a standard specification, whether through national standards or some controlling body, should serve the following purposes:

(a) It should provide test procedures and limits so that a given admixture can be shown to be suitable for the job for which it is intended. Further, it should provide the means by which comparative tests can be made so that decisions can be reached on a sound technical and commercial basis. The testing procedures of this part of the specification will therefore in some way reflect the way that the admixture will be used and will include tests of concrete containing the admixture for workability, water reduction, air content, permeability, etc.
(b) The admixture should be shown to cause no harmful or deleterious effect on the concrete in the short or long term. The test procedures may in this case either be associated with evaluation in cement pastes, e.g. the 'soundness' test in Germany, in concrete, e.g. bleeding, freeze–thaw tests, drying shrinkage, or by chemical or physical tests on the admixture itself, e.g. halogen content.
(c) The admixture should be shown to be consistent from batch to batch to assure the user that the product being used is of uniform quality. This is usually covered by specifying methods for solids content,specific gravity and sometimes infra-red identification spectra. Equally important is a standard procedure for sampling and action to be taken when a dispute occurs. In some cases (e.g. in Germany) the situation is reinforced by random sampling from admixture companies' stocks and examination of quality control records by an independent test institute.
(d) The specification should standardize the essential information which must be provided, including instructions for use, the units to be used, container markings, and storage life and conditions.

6.3 Review of international standards and regulations: Historical background

A tentative standard specification for air-entraining admixtures for concrete [1] was published in 1950 by the American Society for Testing and Materials. In 1962 the first edition of their standard specifications for chemical admixtures appeared [2]. Standard specifications for calcium chloride for use in concreting were published by Spain [3] in 1958 and the UK [4] in 1963. The specifications however were limited to the preparation and use of admixtures, but did not cover performance requirements. The majority of the standard specifications existing today were published in the previous decade. At least twelve countries now possess such standards [1–19] and in several other countries regulations [20–24] governing the approval of the admixture exist. Additionally codes or guides on the use of admixtures were published in Australia [25], Canada [26], Denmark [27], and the United States [28]. In 1975 the RILEM working group produced their final report on the use, classification, quality control and testing of admixtures [29].

6.4 Standards specifications, regulations and codes

The difference in terminology and definition found in the standards of various countries make a standards comparison difficult. Therefore, in the absence of an ISO standard on classification the definitions of admixtures used in the text are those found in the RILEM working group final report [29]. Table 6.1 presents a list of admixtures covered by the standards of a number of countries. The admixtures are listed in three groups according to the main effects they produce in concrete. In certain cases the description of an admixture given in national standards will not necessarily correspond to the RILEM designation. A particular example can be cited in the Austrian standard [6] where a material designated an 'antifreezing' admixture in this standard is classified in the in the RILEM report as accelerator, water-

Table 6.1 Types of admixture covered by national standard specifications reviewed

Types of admixture	Symbols
Water-reducing	WR
Air-entraining	AE
Set-retarding	SRe
Set-accelerating	SAc
Accelerators of hardening*	Ac
Expansion producing	Ex
Antifreezing	aF
Permeability-reducing	PR
Bonding	B
Colouring	C
Water-reducing and air-entraining	WR-AE
Water-reducing and set-retarding	WR-SRe
Water-reducing and set-accelerating	WR-SAc
Water-reducing accelerators of hardening	WR-Ac
Air-entraining accelerators of hardening	AE-Ac
Antifreezing accelerators of hardening	aF-Ac
Permeability-reducing and water-repelling	PR-aW
Strength-increasing	StI
Strength-increasing and set-retarding	StI-SRe
Permeability-reducing and liquid-repelling	PR-aL

*Which very often shorten the setting period.

Table 6.2 Types of admixture covered by standards of individual countries

Country	Type of admixture																			
	WR	AE	SRe	SAc	Ac	Ex	aF	PR	B	C	WR-AE	WR-SRe	WR-SAc	WR-Ac	AE-Ac	aF-Ac	PR-aW	StI	StI-SRe	PR-aL
Australia	X	X	X	X	X							X	X							
Austria					X									X	X					
Belgium	X	X	X	X	X			X			X									
Bulgaria	X	X	X	X	X			X			X	X		X		X				
Canada	X	X	X		X							X						X	X	
Columbia	X		X		X							X		X						
France*	X	X	X	X	X	X	X		X							X	X			X
Israel	X	X	X	X								X	X							
Italy	X	X	X		X						X	X		X		X				
Romania	X																			
United Kingdom	X	X	X		X					X		X		X						
USA	X	X	X		X							X		X						

* Norme Française, NF P 18–103 deals primarily with classification and marking.

reducing accelerator or air-entraining accelerator. Table 6.2 presents a list of admixture types covered by some national standards.

In some instances the regulations governing the approval of admixtures may not be contained in national standards. In Germany the code for the design and construction of concrete structures [30] only permits the use of admixtures having a valid test mark. The directives governing the allocation of test marks [21] and the control [22] and testing [23] of admixtures are issued by the Institut für Bautechnik in Berlin. These directives relate to water-reducing, air-entraining, set-retarding, accelerating and permeability reducing admixtures. Also the French standard NFP 18–103 [12] deals primarily with classification, definitions and marking. The regulations governing the approval of admixtures in France [20] are issued by COPLA (Commission Permanente des Liants Hydrauliques et des Adjuvants du Béton). Comprehensive uniformity and performance requirements are contained in these regulations.

6.5 Uniformity requirements of specifications

A majority of the specifications usually include some chemical and physical tests as a means ensuring uniformity of the products being supplied. Twenty tests suitable for the adequate categorization of an admixture have been identified [29]. These tests are given below:

General:

(a) Physical state: observation of physical nature, appearance and microscopical characteristics

Quantitative determination of specific constituents:

(b) Dry solids content
(c) Loss of ignition
(d) Insoluble residue after ignition
(e) Sugar types and content
(f) Chloride content
(g) Sulphate content
(h) Nitrate and ammonia content
(i) Phosphate content

Specific properties:

(j) pH
(k) Surface tension
(l) Foam stability
(m) Specific surface
(n) Specific mass
(o) Specific volume

(p) Freezing point of solutions
(q) Electrochemical behaviour

Identification:
(r) Infra-red and/or X-ray spectra
(s) X-ray fluorescence
(t) Check on validity of manufacturer's methods of identification.

Most of these properties or compositional features are not of critical importance themselves and therefore, not all the listed tests are appropriate for all admixtures. Most standards specifications stipulate much fewer tests. The most frequently specified properties are items (b), (j) and specific gravity. The RILEM final report suggests only tests (a), (b), (r) or (s) and (t) for quality control purposes. Only two countries' standards (Australia [5] and United States [2]) make provision for similarity of the infra-red spectra of initial and test samples to be checked. These tests are however optional.

Table 6.3 Attributes of admixtures upon which compliance limits are placed by standards and regulations of individual countries

Attribute of admixture	Country											
	Australia	Austria	Belgium	Bulgaria	Canada	France	Germany	Israel	Italy[d]	Romania	UK	USA[g]
†**General*												
Physical state	X	X	X							X		
Content of particular constituents												
†*Dry solids[a]	X		X		X	X		X	X	X	X[f]	X
*Loss on ignition[b]			X							X	X	
Insoluble residue		X	X					X	X[e]	X		
*Insoluble residue after ignition												
Moisture					X[c]							
*Reducing sugars			X			X		X	X	X		
*Chlorides												
Total			X			X	X	X	X		X	
Water soluble		X	X									
Anhydrous calcium chloride		X										
*Sulphates												
Sulphur			X									
*Nitrate and ammonia												
*Phosphate												
Sulphonated lignin									X			

Table 6.3 continued

Attribute of admixture	Country: Australia	Austria	Belgium	Bulgaria	Canada	France	Germany	Israel	Italy[d]	Romania	UK	USA[g]
Special properties												
*pH	X		X		X	X		X	X	X	X	
Total alkalinity									X			
*Surface tension									X			
*Foam stability												
Foam capacity									X[k]			
*Specific surface												
{ Relative density	X		X		X	X	X	X	X			X
{ Specific volume												
Apparent density						X[h]						
*Freezing point			X									
*Electrochemical behaviour							X					
Measurment of colour			X									
Identification												
†* { Infra-red spectra	X											X
†* { X-ray spectra												
†* { X-ray fluorescence												
†* { UV absorption	X											
†* Check on manufacturer's methods												

Notes: *Recommended by RILEM report [29] for adequate characterization of an admixture. †Recommended by RILEM report [29] for control of the admixture itself.

a. Residue at 105 °C; alternatively expressed as weight loss at 105 °C; b. Alternatively expressed as residue on ignition (ash); c. For non-liquid admixtures; d. Tests called for but no limits specified; e. In distilled water and lime saturated water; f. By distilling the admixture in a carrier liquid completely immiscible with and lighter than water, and collecting and measuring the water evolved; g. When requested by purchaser in order to test admixture for uniformity or equivalence; h. For powdered admixtures; k. Only for the admixture type 'water-reducing and air-entraining'.

The relevant admixture properties upon which compliance limits are placed by standards, regulations and codes of practice of different countries are presented in Table 6.3. Individual standards stipulate different combinations of these requirements for different admixtures depending on their type. One of the items worthy of mention is the chloride content. Several countries place compliance limits on the chloride content of admixtures (Table 6.3). There is also an increasing tendency to ban or restrict severely the use of chloride-based accelerators in prestressed and reinforced concrete and in concrete containing embedded metals [6, 30, 31]. The compliance criteria are given in the form of allowable tolerances on

(a) The values obtained for the initial sample [2, 21],
(b) The mean values declared by the manufacturer [13, 18] or a range [17] of the mean values given by the manufacturer.

6.6 Performance requirements of specifications

These requirements deal with the performance of admixtures in cement paste, mortar and concrete. The relative performance of paste mortar and/ or concrete containing an admixture (test mixes) is compared to a corresponding paste, mortar or concrete of similar consistency and containing the same materials but no admixture (reference mixes). The stipulation of performance requirements in paste, mortar or concrete or combinations of these differs in the various standards. This is presented in

Table 6.4 Types of test and reference mix for which compliance limits are imposed by standards and regulations of individual countries

Admixture type	Country											
	Australia	Austria	Belgium	Bulgaria	Canada	France	Germany	Israel	Italy	Romania*	UK	USA
	—	—	P	—	—	—	P	—	P	—	—	—
WR	—	—	—	—	—	M	—	—	M	—	—	—
	C	—	C	C	C	C	C	C	C	—	C	C
	—	—	P	—	—	—	P	—	P	—	—	—
AE	—	—	—	—	—	M	—	—	M	—	—	M†
	C	—	C	—	C	C	C	C	C	—	C	C
	—	—	—	P	—	—	P	—	P	—	—	—
SRe	—	—	M	—	—	M	—	—	M	—	—	—
	C	—	C	C	C	—	C	C	C	—	C	C
	—	P	—	P	—	—	—	—	P	—	—	—
Ac	—	M	—	—	—	M	—	—	M	—	—	—
	C	C	C	C	C	—	C	C	C	—	C	C
	—	—	P	—	—	—	P	—	—	—	—	—
PR	—	—	—	—	—	—	—	—	—	—	—	—
	—	—	C	C	—	—	—	—	—	—	—	—

Note: P, M and C indicate criteria for tests on paste, mortar and concrete respectively. *The Romanian standards [15,16] do not include performance requirements. †Optional.

Table 6.4 which shows the types of mixes used for compliance testing by the standards and regulations of individual countries.

Each specification details the reference mix with particular regard to cement content and consistency. Mean cement contents and slump values ranging from 270 kg m^{-3} [6, 10] to 350 kg m^{-3} [20] and 50 mm [8] to 100 mm [13] respectively have been stipulated. The UK standard specifies consistency in terms of compaction factor while Austrian and Belgium standards quote flow table values.

The choice of constituent materials used in the test and reference mixes is far from uniform in the various standards. Some standards name the source and overall grading for aggregate, and the COPLA technical regulations call for specific cement compositions from named plants. In contrast ASTM-C 494 [2] recommends that except where tests relate to specific uses, a thorough blend of equal parts of three cements from three mills should be used. It specifies the grading of coarse and fine aggregates, but otherwise just requires the materials to be supplied from single lots which conform to ASTM specification for concrete aggregate. The use of a selected 'typical' ordinary Portland cement is specified in the British Standard BS 5075, Part 1 and 2 [18].

The properties upon which limits are placed as well as the test methods employed to measure them differ markedly in the various standards. In individual specifications, however, the performance requirements selected are designed to give assurance that each type of admixture will perform in the required manner and will not adversely effect the concrete properties by, for example, increasing bleeding, final setting time or shrinkage or by reducing strength or durability.

Two inherent problems arise in testing admixture performance.

(a) Variations in the individual materials used to make the concrete tested under laboratory conditions and those used for a specific job.
(b) Differences between laboratory conditions and job conditions.

Interactions between an admixture and differing constituent materials will vary, thus affecting its performance. The provisions made in the British and ASTM standards to reduce this variation are not effective. An admixture which performs satisfactorily with a 'typical' cement or one blended from say, three mills will not necessarily give similar performance when used with a different cement. The RILEM final report suggests that depending upon the country concerned, tests should be made using several cements. Only the French COPLA regulations [20] require the use of more than one cement; they specify that two named cements, one of relatively low C_3A content and one of relatively high C_3A content shall be used. In like manner admixture performance can be affected by aggregates having different properties and grading and by the type of mixing water (hard or soft) used.

Table 6.5 Specification performance requirements relating to paste containing different types of admixture

Property		Country and admixture types concerned						
		Austria	Belgium	Germany	Italy			
		Ac*	WR, AE and PR	WR, AE and SRe	WR	AE	SRe	Ac
Water for standard consistence		—	—	—	≤0.97 R	—	—	—
Setting time h : min	Initial	≥0 : 45 ≤R + 1 h	≥0 : 45	≥1 h	Must comply with standard for cements		≥R + 1 : 30	≥ {Cements standard less 15 min
	Final	≤R + 1 h	≥3 h ≥12 h	<16 h (<12 h if to be used in prestressed concrete)			≥R + 1 : 30	Must comply with standard for cements
Volume stability		Must comply with standard for cements	Determined as for cement: spreading ≤6 mm†	*'Biscuit' test* (i) No cracking (ii) Deflection <2 mm	Expansion in autoclave ≤(R + 0.10)%			

*Described in ÖNORM B 3332 as 'antifreezing'.
†i.e. spreading of indicator points.

ASTM-C 494 [2] recommends that for testing for specific uses, those materials to be used in the concrete for the job should be employed.

Laboratory test conditions are chosen to give standardized testing. Admixture performance on site, however, may be adversely affected by differences in ambient or concrete temperature, workability time of agitation, type and degree of compaction, actual freeze-thaw condition occurring, etc. Nevertheless it must be appreciated that standardized testing in the laboratory according to tests detailed in specifictions chiefly provides criteria for selection of an admixture. Subsequent on-site testing then determines the compatibility of the admixture with materials to be used on the job under job site conditions. Additionally such on-site testing enables the user to ascertain level of admixture dosage required to enhance a desired secondary function such as high workability. Thus although the testing under job site conditions may not corroborate laboratory testing, it provides verification that the admixture selected (from the laboratory testing) will in fact do the job under job site conditions.

6.6.1 Performance requirements in cement paste

The Austrian [6], Belgian [7], Bulgarian [8], Italian [14] and the German Institut für Bautechnik directives [21–23] call for admixture compliance tests to be carried out in cement paste. Properties determined include water requirement to produce a standard consistency, initial and final setting times and volume stability or soundness. The compliance criteria of the above standards as these relate to five admixtures is presented in Table 6.5. Note that in Table 6.5 and in the later Tables 6.6 and 6.7, R represents the value of the property obtained on the reference paste, mortar or concrete.

6.6.2 Performance requirements in mortar

Such testing is required for certain types of admixtures by the Austrian [6], German [21], Belgian [7], Italian [14] and United States standards and by the French COPOLA regulations. The properties specified in these standards include the following:

(a) Water-reduction
(b) Consistency (at same w/c ratio)
(c) Air content
(d) Bleeding
(e) Setting times
(f) Heat of hydration
(g) Strength: compressive and flexural
(h) Shrinkage
(i) Capillary absorption

Table 6.6 Specification requirements relating to mortar containing different types of admixture

Country	Austria	Belgium	Italy				USA
Type of admixture	Ac[a]	SRe	WR	AE	SRe	Ac	AE
Water reduction	—	—	⩾0.06 R	—	—	—	—
Workability (at same w/c ratio)	—	—	⩾1.15 R[b]	—	—	—	—
Air content	—	—	—	⩾(R + 7)%	—	—	Within range (X ± 2)%[c]
Initial set	—	>R + 1 h	—	—	—	—	—
Heat released	>R	—	—	—	—	—	—
Compressive and flexural strength	—	—	1 day: ⩾0.90 R at other ages: ⩾0.95 R	at all ages: ⩾0.60 R	28 day: ⩾0.95 R	1 day: ⩾1.15 R at other ages: ⩾R	—
Shrinkage	—	—	⩽(R + 0.010)%	⩽(R + 0.010)%	⩽(R + 0.010)%	⩽(R + 0.030)%	—

Table 6.6 Continued

Country		France							Germany
Type of admixture		WR		AE	SRe	Ac		PR[g]	PR
Bleeding		—		—	?	—		—	—
		At 5 °C	At 20 °C						
Setting time h : min	Initial	≥R + 1 : 30 ≤R + 3	≥R + 1 ≤R + 2	≥R − 1 ≤R − 2	≥R + 1 ≤R + 3 day	≥0 : 30 ≤R − 1 } f		≥R + 1 ≤R + 2	— —
	Final	≥R + 2 ≤R + 4	≥R + 1 : 30	≥R − 1 : 30 ≤R + 3	Final less initial ≥R	Final less[f] initial ≤R		≥R + 1 : 30 ≤R + 3	— —
Heat of hydration[d]	6 h	—		—	—	≥1.50 R		—	—
	12 h	—		—	—	≥1.30 R		—	—
	1day	—		—	—	≥1.05 R		—	—
	5day	—		—	—	≥R		—	—
						At 5 °C	At 20 °C		
Compressive and flexural strength	1 day	—		—	—	≥0.20 R	≥1.15 R	—	—
	3 day	—		—	—	≥0.50 R	≥R	≥0.85 R	—
	7 day	—		—	≥0.80 R	—	—	≥0.85 R	—
	28 day	—		—	≥0.90 R	—	—	≥0.85 R	—
	90 day	—		—	≥R	—	—	≥0.85 R	—
Shrinkage	7 day		≤1.05 R	—	≤1.25 R	≤1.40 R		≤1.10 R	—
	28 day		≤1.05 R	—	≤1.10 R	≤1.40 R		≤1.10 R	—
	90 day		≤1.05 R	—	≤1.10 R	≤1.30 R		≤1.10 R	—
Capillary absorption	7 day age	—		≤R[e]	—	—		≤0.50 R[e]	≤0.60 R
	90 day age	—		≤R[e]	—	—		≤0.50 R[e]	≤0.60 R

See page 288 for notes to this table.

Table 6.7 Specification performance requirements relating to non air-entrained concrete containing **water-reducing** admixtures

Country		Australia	Belgium	Bulgaria	Canada	France	
						Low C_3A	*High C_3A*
Water reduction		⩾0.05 R	⩾0.05 R	⩾0.05 R	⩾0.05 R	⩾15 litres ⩾0.086 R	⩾12 litres ⩾0.069 R
Consistency (at same w/c)		—	—	—	—	—	
Air content		—	—	—	—	at N ⩽5% at 3N ⩽6% } d	
Bleeding		—	—	—	—	—	
Setting or stiffening time, h : min	Initial	⩾R − 1 ⩽R + 1	—	—	⩾R − 1 : 20 ⩽R + 1 : 20	—	
	Final	⩾R − 1 ⩽R + 1					
Compressive strength	1 day	—	—	—	—	—	
	3 day	⩾1.10 R	⩾1.10 R	—	⩾1.15 R/1.05[a]	⩾1.10 R	
	7 day	⩾1.10 R	—	⩾1:05 R	⩾1.15 R/1.05	⩾1.10 R	
	28 day	⩾1.10 R	⩾1.10 R	⩾1.10 R	⩾1.15 R/1.05	⩾1.10 R	
	90 day	⩾R	—	—	—	⩾1.10 R	
	180 day	⩾R	—	—	⩾R/1.05 } b	—	
	365 day	⩾R	—	—	⩾R/1.05 } b	—	
Tensile strength		Splitting 3, 7, 28 and 90 day ⩾R	—	—	—	Splitting 3, 7, 28 and 90 day ⩾1.05 R	
Shrinkage		—	—	—	(i) ⩽1.35 R[c] (ii) ⩽R + 0.010%	—	

Table 6.7 Continued

Country		Israel	Italy	UK	USA	Germany
Water reduction		⩾0.08 R	⩾0.05 R	⩾0.05 R	⩾0.05 R	Less than control mix
Consistency (at same w/c)		—	⩾1.10 R	—	—	—
Air content		⩽(R + 1)%	—	⩽(R + 2)% and ⩽3%	—	—
Bleeding		Capacity and duration within range R ± D[e]	⩽0.95 R	—	—	—
Setting or stiffening time, h : min	Initial	⩾R − 1	—	⩾R − 1	⩾R − 1	—
		⩽R + 1	—	⩽R + 1	⩽R + 1 : 30	—
	Final	⩾R − 1 } f	—	⩾R − 1	⩾R − 1	—
		⩽R + 1 } f	—	⩽R + 1	⩽R + 1 : 30	—
Compressive strength	1 day	⩾1.10 R	⩾1.05 R	—	—	—
	3 day	⩾1.10 R	⩾1.05 R	—	⩾1.10 R } g	—
	7day	—	⩾1.10 R	⩾1.10 R	⩾1.10 R } g	—
	28 day	⩾1.10 R	⩾1.15 R	⩾1.10 R	⩾1.10 R } g	—
	90 day	⩾1.10 R	⩾1.15 R	—	— } g	—
	180 day	—	—	—	⩾R } g	—
	365 day	—	—	—	⩾R } g	—
Tensile strength		Flexural 28 day ⩾1.05 R	—	—	Flexural 3, 7 and 28 day ⩾R } g	—
Shrinkage		⩽1.20 R at 3 months	—	—	(i) ⩽1.35 R } c	—
					(ii) ⩽R + 0.010% } c	—

See page 288 for notes to this table.

Notes to Table 6.6

a. Actually described by ÖNORM B3332 as 'antifreezing'; b. Measured by flow table test; c. Optional requirement for mortars prepared from successive lots of the admixture. X is the value of air content obtained with the original acceptance sample; d. Using Langavant Calorimeter; e. Capillary absorption of mortar containing admixture must be not greater than limit given after testing for both 1 and 7 days; f. At both 5 °C and 20 °C; g. Actually 'permeability-reducing and water-repelling'.

Table 6.7(a) Specification performance requirements relating to non-air entrained concrete containing **superplasticizers**

Country		USA		Germany
		Non-retarding type	Retarding type	
Water reduction		⩾0.12 R	⩾0.12 R	—
Consistence (at same w/c)		—	—	⩾12 cm increase in flow table test
Setting or stiffening h : min	Initial	⩾R − 1 ⩽R + 1 : 30	⩾R + 1 ⩽R + 3 : 30	— —
	Final	⩾R − 1 ⩽R + 1 : 30	— ⩽R + 3 : 30	— —
Compressive strength	1 day	⩾1.40 R	⩾1.25 R	—
	3 day	⩾1.25 R	⩾1.25 R	—
	7 day	⩾1.15 R	⩾1.15 R	—
	28 day	⩾1.10 R	⩾1.10 R	—
	90 day	—	—	—
	180 day	⩾R	⩾R	—
	365 day	⩾R	⩾R	—
Flexural strength	3 day	⩾1.1R (g)	⩾1.1R (g)	—
	7 day	⩾R (g)	⩾R (g)	—
	28 day	⩾R (g)	⩾R (g)	—
Shrinkage	I.	⩽1.35 R (c)	I. ⩽1.35 R (c)	—
	II.	⩽R + 0.01% (c)	II. ⩽R + 0.01% (c)	—

Notes to Table 6.7. and 6.7(a)

a. 1.05 is a multiplying factor applied to the test mix strength results as an allowance for testing tolerance; b. Only at purchaser request or if no previous test or service record available; c. If R exceeds 0.030%, limit (i) applies; if R is not more than 0.030%, limit (ii) applies; d. N denotes normal recommended dose of WR; 3N denotes 3 × normal recommended dose of WR; e. Where D is the acceptable deviation quoted by the manufacturer, if declared; f. Not obligatory; g. No strength test result for the test mix must be less than 90% of that at any prior age; ie. it is required that the strength of the test mix does not decrease with age.

Table 6.6 presents the differing selection of compliance criteria specified by the various standards.

6.6.3 Performance requirements in concrete

Most standard specifications now require admixtures to meet performance requirements in concrete. The determination of the following characteristics is usually called for:

(a) Water reduction or water content
(b) Consistency (at same w/c ratio)
(c) Air content
(d) Bleeding
(e) Setting or stiffening time
(f) Compressive strength
(g) Tensile strength
(h) Shrinkage
(i) Freeze–thaw resistance

The performance criteria required for water-reducing, air-entraining, set-retarding accelerators and permeability-reducing admixtures are indicated in Tables 6.7 to 6.11 respectively.

6.7 Information to be provided by the manufacturer or vendor

The information given below is supplied by the manufacturer to ensure that an admixture is properly identified, handled, stored and used correctly and safely. Some standards specify in great detail the information which the manufacturer/vendor must supply. The types of information that the standards either demand, or allow the purchaser to require, of the manufacturer/vendor are as follows:

(a) Identification of admixture by name, type, etc.
(b) Details of acceptance and uniformity tests
(c) Details of any certification, i.e. details of conformance to standard
(d) Recommended normal and maximum dosage
(e) Effects of under-dosage and over-dosage
(f) Secondary effects of the admixture on air content, setting/stiffening, or colour of concrete
(g) Instructions for transport and storage
(h) Instructions for use including any restrictions on use and any necessary safety precautions
(i) Physical state

Table 6.8 Specification performance requirements relating to concrete containing **air-entraining** admixtures

Country		Australia[a]	Belgium	Bulgaria	Canada[a]	France
Water content		—	—	≥0.95 R ≤1.10 R	—	—
Air content		—	≥(R + 2.5)%	—	—	at N, ≥ (R + 2)% at M, ≤ (R + 4)% } k
Bleeding		≤(R + 2)%	—	—	—	—
Setting or stiffening time, h : min	Initial	≥R − 1 ≤R + 1	— —	— —	≥R − 1 ≤R + 1	— —
	Final	≥R − 1 ≤R + 1	—	—	—	—
Compressive strength	1 day	—	—	≥0.90 R (age unspecified)	—	—
	3 day	—	$\geq R - 4\ A_t$ } b		≥R/1.05[c]	≥0.80 R
	7 day	≥0.90 R	—		≥R/1.05	≥0.80 R
	28 day	≥0.90 R	$\geq R - 4\ A_t$ } b		≥R/1.05	≥0.80 R
	90 day	≥0.90 R	—		—	≥0.80 R
	180 day	—	—		≥R/1.05 } d	—
	365 day	≥0.90 R	—		≥R/1.05 } d	—
Tensile strength		—	—	—	—	Splitting 3, 7, 28 and 90 day ≥0.80 R
Shrinkage		—	—	—	(i) ≤1.20 R (ii) ≤R + 0.006% } e	—
Freeze–thaw resistance		—	—	≥R + '1 grade'	AVSF ≤ 200 μm or RDF ≥R/1.10 } f,g	(i) Strength ≥0.70 R (ii) Ed ≥50% } m

Table 6.8 Continued

Country		Israel	Italy	UK	USA[a]	Germany
Water content		≤0.92 R	—	—	—	—
Air content		≥(R + 4)% ≤(R + 6)%	≥(R + 3)%	Repeatability[r] 4% ≤ A ≤6%	—	≥(R + 3.5) ≤(R + 4)
Bleeding		Capacity and duration within range R ± D[n]	≤(R − 5)%	—	≤(R + 2)%	—
Setting or stiffening time, h : min	Initial	≥R − 1 ≤R + 1 : 30	— —	≥R − 1 ≤R + 1	≥R − 1 : 15 ≤R + 1 : 15	— —
	Final	≥R − 1 ≤R + 1 : 30 } p	— —	≥R − 1 ≤R + 1	≥R − 1 : 15 ≤R + 1 : 15	— —
Compressive strength	1 day	≥0.75 R	≥0.90 R at all ages	—	—	—
	3 day	≥0.85 R		—	—	—
	7 day	—		—	—	—
	28 day	≥0.85 R		≥0.70 R	≥0.90 R at any age	—
	90 day	≥0.85 R		—		—
	180 day	—		—	—	—
	365 day	—		—	—	—
Tensile strength		Flexural 28 day ≥0.85 R	—	—	≥0.90 R[t] at any age	—
Shrinkage		≤1.20 R at 3 months	—	—	(i) ≤1.20 R (ii) ≤R + 0.006% } e,t	—
Freeze-thaw resistance		—	DF ≥1.30 R[q]	$\frac{l_{50} - l_0}{L_0}$ ≤0.0005[s]	RDF ≥80%[u]	—

See page 296 for notes to this table.

Table 6.9 Specification performance requirements relating to non-air-entrained concrete containing **set-retarding** admixtures

Country		Australia	Belgium	Bulgaria	Canada
Water content		—	—	—	≤0.97 R
Air content		—	—	—	—
Bleeding		—	—	—	—
Setting or stiffening time, h : min	Initial	≥R + 1	—	—	≥R + 1
		≤R + 3	—	—	≤R + 3
	Final	≥R + 1	—	—	—
		≤R + 3	—	—	—
Compressive strength	1 day	—	—	—	—
	3 day	≥0.90 R	≥0.90 R[a]	—	≥1.15 R/1.05[b]
	7 day	≥0.90 R	—	—	≥1.15 R/1.05
	28 day	≥R	≥R[a]	≥R	≥1.15 R/1.05
	90 day	≥R	—	—	—
	180 day	≥R	—	—	≥R/1.05 } c
	365 day	≥R	—	—	≥R/1.05 } c
Tensile strength		Splitting 3 and 7 day ≥0.90 R; 28 and 90 day >R	—	—	—
Shrinkage		—	—	—	(i) ≤1.35 R; (ii) ≤R + 0.010% } d

Table 6.9 Continued

Country		France[e]	Israel	Italy	UK	USA
Water content		—	—	—	—	—
Air content		—	⩽(R + 1)%	—	⩽(R + 2)% and ⩽3%	—
Bleeding		—	Capacity and duration within range R ± D[f]	—	—	—
Setting or stiffening time, h : min	Initial	—	⩾R + 1 : 30	⩾R + 0 : 45	⩾R + 1	⩾R + 1 ⩽R + 3 : 30
	Final	—	⩽R + 4	⩾R + 0 : 45	—	⩽R + 3 : 30
Compressive strength	1 day	—	—	—	—	—
	3 day	—	⩾0.90 R	—	—	⩾0.90 R [g]
	7 day	—	⩾R	—	⩾0.90 R	⩾0.90 R [g]
	28 day	—	⩾R	—	⩾0.95 R	⩾0.90 R [g]
	90 day	—	⩾R	⩾R	—	⩾0.90 R [g]
	180 day	—	—	—	—	⩾0.90 R[h] [g]
	365 day	—	—	—	—	⩾0.90 R[h] [g]
Tensile strength		—	Flexural 28 day ⩾0.90 R	—	—	Flexural 3, 7 and 28 day ⩾0.90 R [g]
Shrinkage		—	⩽1.20 R at 3 months	—	—	(i) ⩽1.35 R (ii) ⩽R + 0.010% [d]

See page 296 for notes to this table.

Table 6.10 Specification performance requirements relating to non-air-entrained concrete containing **accelerators of hardening**

Country		Australia	Austria	Belgium	Bulgaria	Canada
Water content		—	—	—	—	≤R
Air content		—	—	—	—	—
Bleeding		—	—	—	—	—
Setting or stiffening time, h : min	Initial	≥R − 3 ≤R − 1	—	—	—	≥R − 1 ≤R − 3
	Final	≥R − 3 ≤R − 1	—	—	—	—
Compressive strength	1 day	—	—	At 20 °C ≥1.25 R	≥1.15 R	—
	3 day	≥1.25 R	—	At 5 °C ≥1.25 R	≥1.15 R	≥1.25 R/1.05[a]
	7 day	≥R	—	—	—	≥1.10 R/1.05
	28 day	≥R	≥0.90 R	At 5 ° and 20 °C ≥R	≥1.10 R	≥R
	90 day	≥R	—	—	—	—
	180 day	—	—	—	—	≥R/1.05 } b
	365 day	≥R	—	—	—	≥R/1.05 } b
Tensile strength		Splitting 3 day ≥1.10 R 7, 28 and 90 day ≥R	—	—	—	—
Shrinkage		—	≤2 R at 90 days	—	—	(i) ≤1.35 R (ii) ≤R + 0.010% } c

Table 6.10 Continued

Country		France[d]	Israel	Italy	UK	USA
Water content		—	—	—	—	—
Air content		—	≤(R + 1)%	—	≤(R + 2)% and <3%	—
Bleeding		—	Capacity and duration within range R ± D[e]	—	—	—
Setting or stiffening time, h : min	Initial	—	≥R − 3 : 30 ≤R − 1	≤R − 0 : 30	>1	≤R − 1 ≥R − 3 : 30
	Final	—	≥R − 3 : 30 ≤R − 1	≤R − 0 : 30	≥R − 1	≤R − 1
Compressive strength	1 day	—	≥1.50 R	≥1.15 R	≥1.25 R	—
	3 day	—	≥1.25 R	≥1.15 R	—	≥1.25 R
	7 day	—	—	Also	—	≥R
	28 day	—	≥R	≥R at	≥0.95 R	≥R [f]
	90 day	—	≥R	later standard	—	—
	180 day	—	—	ages of test	—	≥0.90 R[g]
	365 day	—	—	for cement used	—	≥0.90 R[g]
Tensile strength		—	Flexural 28 day ≥R	—	—	Flexural 3 day ≥1.10 R 7 day ≥R 28 day ≥0.90 R [f]
Shrinkage		—	≤1.20 R at 90 days	—	—	(i) ≤1.35 R (ii) ≤R + 0.010% [c]

See page 296 for notes to this table.

Notes to Table 6.8

a. Reference mix contains a reference air-entraining agent (neutralized Vinsol resin); b. A is the additional air content, %, in excess of that of the reference mix; c. 1.05 is a multiplying factor applied to the test mix strength results as an allowance for testing tolerance; d. Only at purchaser's request or if no previous test or service record available; e. If R exceeds 0.030%, limit (i) applies; if R is not more than 0.030%, limit (ii) applies; f. At purchaser's option either air–void spacing factor (ASTM C457) or relative durability factor (ASTM C666, Procedure A) may be used to determine durability; g. 1.10 is a correction factor included as an allowance for testing tolerance; k. N denotes normal recommended dose of AE; M denotes maximum recommended dose of AE; m. After 90 cycles between −16 °C and 5 °C, (i) test mix flexural and compressive strengths are compared with those of test mix prisms not subjected to freeze–thaw cycles; (ii) dynamic modulus is compared with initial dynamic modulus before freezing and thawing; n. Where D is the acceptable deviation quoted by the manufacturer, if declared; p. Not obligatory; q. To UNI 7087–72; r. 'A' indicates individual air contents for three consecutive test mix batches determined by one operator using one set of equipment; s. Relative length change after 50 cycles between −15 °C and 16 °C; t. Applicable only when specifically required by purchaser for structures where flexural strength or volume change may be critically important; u. Relative durability factor of test mix after freeze-thaw cycling from −17.8 °C to 4.4 °C (ASTM C666).

Notes to Table 6.9

a. At both 20 °C and 30 °C; b. 1.05 is a multiplying factor applied to the test mix strength results as an allowance for testing tolerance; c. Only at purchaser's request, or if no previous test or service record available; d. If R exceeds 0.030%, limit (i) applies; if R is not more than 0.030%, limit (ii) applies; e. COPLA criteria for approval of set-retarding admixtures do not include performance requirements for concrete test mixes; f. Where D is the acceptable deviation quoted by the manufacturer, if declared; g. No strength test result for the test mix must be less than 90% of that at any prior age; i.e. it is required that the strength of the test mix does not decrease with age; h. When the admixture is tested for use in specific work the tests at 6 months and 1 year may be waived.

Notes to Table 6.10

a. 1.05 is a multiplying factor applied to the test mix strength results as an allowance for testing tolerance; b. Only at purchaser's request or if no previous test or service record available; c. If R exceeds 0.030%, limit (i) applies; if R is not more than 0.030%, limit (ii) applies; d. COPLA criteria for approval of accelerators of hardening do not include performance requirements for concrete test mixes; e. Where D is the acceptable deviation quoted by the manufacturer, if declared; f. No strength test result for the test mix must be less than 90% of that at any prior age; i.e. it is required that the strength of the test mix does not decrease with age; g. When the admixture is tested for use in specific work, the tests at 6 months and 1 year may be waived.

Table 6.11 Specification performance requirements relating to concrete containing **permeability-reducing** admixtures

Country	Belgium	Bulgaria	France[a]
Water reduction	—	⩾0.05 R	—
Air content	—	—	at N, ⩽(R + 6)%; at M or 3N, ⩽(R + 8)% [b]
Strength	⩾0.90 R at any age	—	—
Watertightness	—	⩾R + 10^5pa	

Notes: a. The admixture type described as 'Hydrofuge' in Norme Française NF P 18–103 is 'permeability-reducing and water-repelling'; b. N denotes manufacturer's normal recommended dose; M denotes manufacturer's maximum recommended dose.

(j) Composition including type of main active ingredient
(k) Chloride content
(l) Sugar content
(m) Any known incompatibility with other concrete constituents

6.8 Labelling

A number of standards [2, 6, 10, 12, 13, 16, 18] contain stipulations about labelling. Packaging is required to carry a label containing appropriate items of information from the above list. Some standards also require several of the following items to be marked:

(a) Quantity
(b) Identity of production lot (batch number, date and location of manufacturer)
(c) Maximum storage life
(d) Endorsement of conformity to standard as approval
(e) Number of relevant standard

6.9 Sampling and testing variability

The producer and consumer risks involved in applying the compliance criteria given in the specifications reviewed are affected by the magnitude of the testing variability. Some allowance for testing variability is explicitly made in the Canadian standards [9, 10] and in the COPLA regulations [20]. The 1979 edition of ASTM C494 [2] contains precision statements for two types of uniformity tests. Elsewhere some allowance for testing variability is thought to be implicit in the rather low level of relative performance required of test mixes. As an example of this, the 5% water reduction generally required of test mixes containing water-reducing admixtures may be quoted.

6.10 Cost

The cost of testing that stems from specifying comprehensive performance requirements [20] should be balanced against the benefit of increased assurance.

References

1 American Society for Testing and Materials. (1979). *ANSI/ASTM C 260–77: Standard specifications for air-entraining admixtures for concrete.* Annual book of ASTM Standards, **Part 14.** Philadelphia, 182–5.

2 American Society for Testing and Materials. (1979). *ANSI/ASTM C 494–79: Standard specification for chemical admixtures for concrete.* Annual book of ASTM Standards, **Part 14,** Philadelphia, 302–12.

3 Instituto Nacional de Racionalizatión del Trabajo. (1958). *UNE 41 113: Cloruro cálcico, utilizado como producto de adición en los hormigones.* (Calcium chloride for use as a concrete admixture). Madrid. 1 pp.

4 British Standards Institution. (1963). *BS 3587: Specifications for calcium chloride.* London, 9 pp. (With Amendment No. 1, 1972).

5 Standards Association of Australia. (1973). *AS 1478: Chemical admixtures for use in concrete.* Sydney, 20 pp. (With Amendment No. 1, May 1974 and Amendment No.2, July 1978.)

6 Österreichisches Normungsinstitut. (1971). *ÖNORM B 3332: Zusatzmittel für Mörtel und Beton. Frostschutzmittel.* (Admixtures for mortar and concrete. Antifreezing agents). Vienna, 7 pp.

7 Institut Belge de Normalisation. (1979). *NBN T61–101: Adjuvants pour mortiers et bétons. Spécifications pour les réducteurs d'eau, entraîneurs d'air, réducteurs d'eau-entraîneurs d'air, réducteurs de perméabilité, retardateurs de prise, accélérateurs de prise et accélérateurs de durcissement.* (Admixtures for mortar and concrete. Specifications for water-reducers, air-entrainers, permeability-reducers, set-retarders, set-accelerators and accelerators of hardening.) Brussels, 7 pp.

8 Bulgarian State Standardisation Committee. (1977). *BSS 14069–77. Concrete admixtures. Classification and technical requirements.* 4 pp. (in Bulgarian).

9 Canadian Standards Association. (1978). *CAN 3 – A 266.1 – M 78. Air-entraining admixtures for concrete.* Rexdale, 19 pp.

10 Canadian Standards Association. (1978). *CAN 3 – A 266.2 – M 78. Chemical admixtures for concrete.* Rexdale, 23 pp.

11 Instituto Columbiano de Normas Tecnicas. (1978). *INCONTEC 1299: Aditivos quimocos para hormigón* (Chemical admixtures for concrete.) Bogota, 9 pp.

12 Association Française de Normalisation. (1972). *NFP 18–103: 72: Définitions et marquage des adjuvants du béton.* (Definitions and marking of concrete admixtures.) Paris, 5 pp.

13 Standards Institution of Israel. (1976). *SI 896: Chemical admixtures for concrete.* (Translation without guarantee.) Tel Aviv, 11 pp.

14 Ente Nazionale Italiano di Unificazione. (1972). *UNI 7101 to 7109–72: Additiviper impasti cementizi.* (Admixtures for cement mixes.) Milan.
15 Institutul Roman de Standardizare. (1970). *STAS 8625–70. Aditiv plastifiant mixt pentru betoane.* (Mixed plasticizing admixture for concrete.) Bucharest, 7 pp.
16 Institutul Roman de Standardizare. (1970). *STAS 8626–70. Lignosulfonat de Calciu tehnic.* (Calcium lignosulphonate, technical grade.) Bucharest, 6 pp.
17 British Standards Institution. (1975) *BS 1014: Specification for pigments for Portland cement and Portland cement products.* London, 11 pp.
18 British Standards Institution. (1982). *BS 5075: Specification for concrete admixtures. Part 1. Accelerating admixtures, retarding admixtures and water-reducing admixtures.* London, 17 pp.
19 British Standards Institution. (1982). *BS 5075: Part 2: Air-entraining admixtures.* London.
20 Commission Permanente des Liants Hydrauliques et des Adjuvants du Béton. (1978). *Règlement technique de l'agrément des adjuvants et des ajouts des bétons.* (Technical regulations for the approval of concrete admixtures and supplementary materials.) Ministère de L'Equipement, Paris.
21 Institut für Bautechnik. (1973). *Richtlinien für die Zuteilung von Prüfzeichen für Betonzusatzmittol.* (Directives for the allocation of test marks for concrete admixtures.) Mitteilungen, **Heft 3,** 86–8. Wilhelm Ernst und Sohn, Berlin.
22 Institut für Bautechnik. (1973). Richtlinien für die Überwachung von Betonzusatzmitteln. (Directives for the control of concrete admixtures.) Mitteilungen, **Heft 3,** 88–90, Wilhelm Ernst und Sohn, Berlin.
23 Institut für Bautechnik. (1974) *Richtlinien für die Prüfung der Wirksamkeit von Betonzusatzmitteln.* (Directives for testing the efficiency of concrete admixtures.) Mitteilungen, **Heft 1,** 1975. 19–22, Wilhelm Ernst und Sohn, Berlin.
24 Suomen Rakennusinsinöörien Liitto Finlands byggnadsingenjörers förbund. (1969). *Betonin lisäaineiden käyttö.* (Use of concrete admixtures.) Helsinki, Suomen Betoniyhdistys r.y., 41 pp.
25 Standards Association of Australia. (1973). *AS 1479: Code of practice for the use of chemical admixtures in concrete.* Sydney, 21–35.
26 Canadian Standards Association. (1978). *CAN3: A 266.4: M 78. Guidelines for the use of admixtures in concrete.* Rexdale, 43.
27 Dansk Ingeniørforenings. (1973). *Anvisning i brug af tilsaetningsstoffer til beton.* (Advice on the use of admixtures in structural concrete.) Copenhagen.
28 American Concrete Institute, Committee 212. (1971). *Guide for use of admixtures in concrete.* ACI Proceedings, **68,** No. 9, 646–76.
29 RILEM Working Group 'Concrete Admixtures'. (1975). *Final Report. Materials and Structures.* November/December, No. 48, 451–72.
30 Deutsches Institut für Normung e.V. (1978). *DIN 1045. Beton und Stahlbetonbau; Bemessung und Ausführung.* (Concrete and reinforced concrete structures. Design and construction.) Berlin, Beuth Verlag Gmbh.
31 British Standards Institution. (1972). *CP 110: The structural use of concrete. Part 1. Design, materials and workmanship.* 154 pp. (With Amendment AMD 2289 of May 1977.)

Index